Runway Incursions

The McGraw-Hill *Controlling Pilot Error* Series

Weather
Terry T. Lankford

Communications
Paul E. Illman

Automation
Vladimir Risukhin

Controlled Flight into Terrain (CFIT/CFTT)
Daryl R. Smith

Training and Instruction
David A. Frazier

Checklists and Compliance
Thomas P. Turner

Maintenance and Mechanics
Larry W. Reithmaier

Situational Awareness
Paul A. Craig

Fatigue
James C. Miller

Culture, Environment, and CRM
Tony Kern

Cover Photo Credits (top row, from left): PhotoDisc; Corbis Images; from *Spin Management and Recovery* by Michael C. Love; PhotoDisc; PhotoDisc; (middle row from left): PhotoDisc; *Piloting Basics Handbook* by Bjork, courtesy of McGraw-Hill; © 2001 Mike Fizer, all rights reserved; image by Kelly Parr; PhotoDisc; (bottom row, from left): © 2002 Mike Fizer, all rights reserved; © 2002 Mike Fizer, all rights reserved.

CONTROLLING PILOT ERROR

Runway Incursions

Bill Clarke

McGraw-Hill

New York Chicago San Francisco Lisbon London Madrid
Mexico City Milan New Delhi San Juan Seoul
Singapore Sydney Toronto

Cataloging-in-Publication Data is on file with the Library of Congress

McGraw-Hill

A Division of The ***McGraw-Hill*** *Companies*

1 2 3 4 5 6 7 8 9 0 DOC/DOC 0 7 6 5 4 3 2

ISBN 0-07-138506-1

The sponsoring editor for this book was Shelley Ingram Carr, the editing supervisor was Steven Melvin, and the production supervisor was Pamela Pelton. It was set in Garamond per the TAB3A design by Deirdre Sheean.

Printed and bound by R. R. Donnelley & Sons Company.

This book is printed on recycled, acid-free paper containing a minimum of 50% recycled, de-inked fiber.

Contents

Series Introduction

The Human Condition

The Roman philosopher Cicero may have been the first to record the much-quoted phrase "to err is human." Since that time, for nearly 2000 years, the malady of human error has played out in triumph and tragedy. It has been the subject of countless doctoral dissertations, books, and, more recently, television documentaries such as "History's Greatest Military Blunders." Aviation is not exempt from this scrutiny, as evidenced by the excellent Learning Channel documentary "Blame the Pilot" or the NOVA special "Why Planes Crash," featuring John Nance. Indeed, error is so prevalent throughout history that our flaws have become associated with our very being, hence the phrase *the human condition.*

The Purpose of This Series

Simply stated, the purpose of Controlling Pilot Error series is to address the so-called human condition, improve performance in aviation, and, in so doing, save a few lives. It is not our intent to rehash the work of over a millennia of expert and amateur opinions but rather to *apply* some of the more important and insightful theoretical perspectives to the life and death arena of manned flight. To the best of my knowledge, no effort of this magnitude has ever been attempted in aviation, or anywhere else for that matter. What follows is an extraordinary combination of why, what, and how to avoid and control error in aviation.

Because most pilots are practical people at heart—many of whom like to spin a yarn over a cold lager—we will apply this wisdom to the daily flight environment, using a case study approach. The vast majority of the case studies you will read are taken directly from aviators who have made mistakes (or have been victimized by the mistakes of others) and survived to tell about it. Further to their credit, they have reported these events via the anonymous Aviation Safety Reporting System (ASRS), an outstanding program that provides a wealth of extremely useful and *usable* data to those who seek to make the skies a safer place.

A Brief Word about the ASRS

The ASRS was established in 1975 under a Memorandum of Agreement between the Federal Aviation Administration (FAA) and the National Aeronautics and Space Administration (NASA). According to the official ASRS web site, *http://asrs.arc.nasa.gov*

> The ASRS collects, analyzes, and responds to voluntarily submitted aviation safety incident reports in order to lessen the likelihood of aviation accidents. ASRS data are used to:
>
> - Identify deficiencies and discrepancies in the National Aviation System (NAS) so that these can be remedied by appropriate authorities.

- Support policy formulation and planning for, and improvements to, the NAS.
- Strengthen the foundation of aviation human factors safety research. This is particularly important since it is generally conceded *that over two-thirds of all aviation accidents and incidents have their roots in human performance errors* (emphasis added).

Certain types of analyses have already been done to the ASRS data to produce "data sets," or prepackaged groups of reports that have been screened "for the relevance to the topic description" (ASRS web site). These data sets serve as the foundation of our Controlling Pilot Error project. The data come *from* practitioners and are *for* practitioners.

The Great Debate

The title for this series was selected after much discussion and considerable debate. This is because many aviation professionals disagree about what should be done about the problem of pilot error. The debate is basically three sided. On one side are those who say we should seek any and all available means to *eliminate* human error from the cockpit. This effort takes on two forms. The first approach, backed by considerable capitalistic enthusiasm, is to automate human error out of the system. Literally billions of dollars are spent on so-called human-aiding technologies, high-tech systems such as the Ground Proximity Warning System (GPWS) and the Traffic Alert and Collision Avoidance System (TCAS). Although these systems have undoubtedly made the skies safer, some argue that they have made the pilot more complacent and dependent on the automation, creating an entirely new set of pilot errors. Already the automation enthusiasts are seeking robotic answers for this new challenge. Not surprisingly, many pilot trainers see the problem from a slightly different angle.

Another branch on the "eliminate error" side of the debate argues for higher training and education standards, more accountability, and better screening. This group (of which I count myself a member) argues that some industries (but not yet ours) simply don't make serious errors, or at least the errors are so infrequent that they are statistically nonexistent. This group asks, "How many errors should we allow those who handle nuclear weapons or highly dangerous viruses like Ebola or anthrax?" The group cites research on high-reliability organizations (HROs) and believes that aviation needs to be molded into the HRO mentality. (For more on high-reliability organizations, see *Culture, Environment, and CRM* in this series.) As you might expect, many status quo aviators don't warm quickly to these ideas for more education, training, and accountability—and point to their excellent safety records to say such efforts are not needed. They recommend a different approach, one where no one is really at fault.

On the far opposite side of the debate lie those who argue for "blameless cultures" and "error-tolerant systems." This group agrees with Cicero that "to err is human" and advocates "error-management," a concept that prepares pilots to recognize and "trap" error before it can build upon itself into a mishap chain of events. The group feels that training should be focused on primarily error mitigation rather than (or, in some cases, in addition to) error prevention.

Falling somewhere between these two extremes are two less-radical but still opposing ideas. The first approach is designed to prevent a reoccurring error. It goes something like this: "Pilot X did this or that and it led to a mishap, so don't do what Pilot X did." Regulators are particularly fond of this approach, and they attempt to regulate the last mishap out of future existence. These so-called rules written in blood provide the traditionalist with plenty of training materials and even come with ready-made case studies—the mishap that precipitated the rule.

Opponents to this "last mishap" philosophy argue for a more positive approach, one where we educate and train *toward* a complete set of known and valid competencies (positive behaviors) instead of seeking to eliminate negative behaviors. This group argues that the professional airmanship potential of the vast majority of our aviators is seldom approached—let alone realized. This was the subject of an earlier McGraw-Hill release, *Redefining Airmanship*.[1]

Who's Right? Who's Wrong? Who Cares?

It's not about *who's* right, but rather *what's* right. Taking the philosophy that there is value in all sides of a debate, the Controlling Pilot Error series is the first truly comprehensive approach to pilot error. By taking a unique "before-during-after" approach and using modern-era case studies, 11 authors—each an expert in the subject at hand—methodically attack the problem of pilot error from several angles. First, they focus on error prevention by taking a case study and showing how preemptive education and training, applied to planning and execution, could have avoided the error entirely. Second, the authors apply error management principles to the case study to show how a mistake could have been (or was) mitigated after it was made. Finally, the case study participants are treated to a thorough "debrief," where alternatives are discussed to prevent a reoccurrence of the error. By analyzing the conditions before, during, and after each case study, we hope to combine the best of all areas of the error-prevention debate.

A Word on Authors and Format

Topics and authors for this series were carefully analyzed and hand-picked. As mentioned earlier, the topics were taken from preculled data sets and selected for their relevance by NASA-Ames scientists. The authors were chosen for their interest and expertise in the given topic area. Some are experienced authors and researchers, but, more important, *all* are highly experienced in the aviation field about which they are writing. In a word, they are practitioners and have "been there and done that" as it relates to their particular topic.

In many cases, the authors have chosen to expand on the ASRS reports with case studies from a variety of sources, including their own experience. Although Controlling Pilot Error is designed as a comprehensive series, the reader should not expect complete uniformity of format or analytical approach. Each author has brought his own unique style and strengths to bear on the problem at hand. For this reason, each volume in the series can be used as a stand-alone reference or as a part of a complete library of common pilot error materials.

Although there are nearly as many ways to view pilot error as there are to make them, all authors were familiarized with what I personally believe should be the industry standard for the analysis of human error in aviation. The Human Factors Analysis and Classification System (HFACS) builds upon the groundbreaking and seminal work of James Reason to identify and organize human error into distinct and extremely useful subcategories. Scott Shappell and Doug Wiegmann completed the picture of error and error resistance by identifying common fail points in organizations and individuals. The following overview of this outstanding guide[2] to understanding pilot error is adapted from a United States Navy mishap investigation presentation.

> Simply writing off aviation mishaps to "aircrew error" is a simplistic, if not naive, approach to mishap causation. After all, it is well established that mishaps cannot be attributed to a single cause, or in most instances, even a single individual. Rather, accidents are the end result of a myriad of latent and active failures, only the last of which are the unsafe acts of the aircrew.
>
> As described by Reason,[3] active failures are the actions or inactions of operators that are believed to cause the accident. Traditionally referred to as "pilot error," they are the last "unsafe acts" committed by aircrew, often with immediate and tragic consequences. For example, forgetting to lower the landing gear before touch down or hotdogging through a box canyon will yield relatively immediate, and potentially grave, consequences.
>
> In contrast, latent failures are errors committed by individuals within the supervisory chain of command that effect the tragic sequence of events characteristic of an accident. For example, it is not difficult to understand how tasking aviators at the expense of quality crew rest can lead to fatigue and ultimately errors (active failures) in the cockpit. Viewed from this perspective then, the unsafe acts of aircrew are the end

result of a long chain of causes whose roots originate in other parts (often the upper echelons) of the organization. The problem is that these latent failures may lie dormant or undetected for hours, days, weeks, or longer until one day they bite the unsuspecting aircrew....

What makes [Reason's] "Swiss Cheese" model particularly useful in any investigation of pilot error is that it forces investigators to address latent failures within the causal sequence of events as well. For instance, latent failures such as fatigue, complacency, illness, and the loss of situational awareness all effect performance but can be overlooked by investigators with even the best of intentions. These particular latent failures are described within the context of the "Swiss Cheese" model as preconditions for unsafe acts. Likewise, unsafe supervisory practices can promote unsafe conditions within operators and ultimately unsafe acts will occur. Regardless, whenever a mishap does occur, the crew naturally bears a great deal of the responsibility and must be held accountable. However, in many instances, the latent failures at the supervisory level were equally, if not more, responsible for the mishap. In a sense, the crew was set up for failure....

But the "Swiss Cheese" model doesn't stop at the supervisory levels either; the organization itself can impact performance at all levels. For instance, in times of fiscal austerity funding is often cut, and as a result, training and flight time are curtailed. Supervisors are therefore left with tasking "non-proficient" aviators with

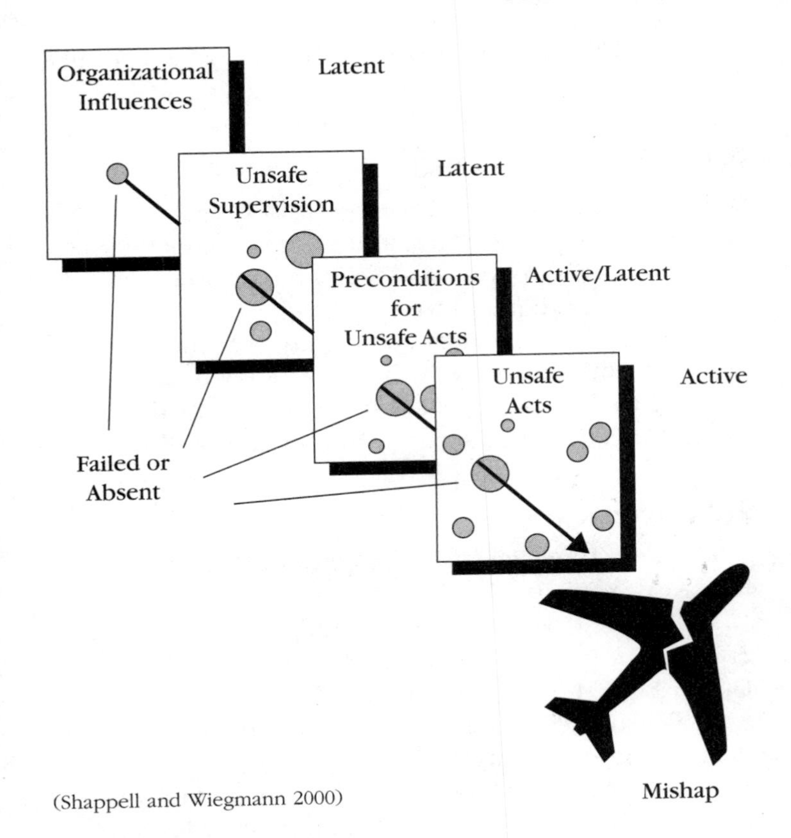

(Shappell and Wiegmann 2000)

> sometimes-complex missions. Not surprisingly, causal factors such as task saturation and the loss of situational awareness will begin to appear and consequently performance in the cockpit will suffer. As such, causal factors at all levels must be addressed if any mishap investigation and prevention system is going to work.[4]

The HFACS serves as a reference for error interpretation throughout this series, and we gratefully acknowledge the works of Drs. Reason, Shappell, and Wiegmann in this effort.

No Time to Lose

So let us begin a journey together toward greater knowledge, improved awareness, and safer skies. Pick up any volume in this series and begin the process of self-analysis that is required for significant personal or organizational change. The complexity of the aviation environment demands a foundation of solid airmanship and a healthy, positive approach to combating pilot error. We believe this series will help you on this quest.

References

1. Kern, Tony, *Redefining Airmanship,* McGraw-Hill, New York, 1997.
2. Shappell, S. A., and Wiegmann, D. A., *The Human Factors Analysis and Classification System—HFACS,* DOT/FAA/AM-00/7, February 2000.
3. Reason, J. T., Human Error, Cambridge University Press, Cambridge, England, 1990.
4. U.S. Navy, *A Human Error Approach to Accident Investigation,* OPNAV 3750.6R, Appendix O, 2000.

Foreword

From many years ago, at an outdoor survival briefing I took as part of an "Outward Bound" adventure course, I recall the words of a young and misguided instructor, "The only certain way to avoid becoming lost, is to never leave the marked trail." It sounded good at the time, perfectly logical. I mean, how can you get lost if you never leave a clearly marked trail? Just follow one signpost to the next, right? Over the ensuing years, I came to learn how wrong he was.

In the decades since my "Outward Bound" experience, I have been seriously disoriented (real pilots don't say "lost") on any number of "marked trails." Thankfully, this has occurred most often in my Ford F-250 on backcountry roads, but occasionally when driving on clearly marked highways and interstates. Usually, the worst that happens when this occurs is that we have to sacrifice our pride and stop and ask for instructions. It is a humbling experience, and there seems to be something in the "Y chromosome that forces us to burn up at least a half a tank of gas before we do so.

On foot or in an automobile, this tendency might cause us some personal frustration or perhaps to be late to our destination. But when this occurs in an aircraft at a crowded aviation terminal, the results can be far more tragic.

In October of 2001, one such disaster occurred when the pilot of a small, privately owned Cessna aircraft in Milan, Italy mistakenly followed a "clearly marked path" onto an active runway at a major international airport. The weather was poor, visibility was limited, and for reasons we may never know, the pilot made a turn into the path of an SAS MD-87 airliner with 104 passengers and six crewmembers on board. The big jet hit the Cessna, and then swerved into a baggage handling depot where it burst into flames, killing four more. It was an entirely preventable mishap, and one that, if we do not confront the problem of runway incursions, will undoubtedly reoccur.

Although this mishap occurred under conditions of poor visibility, nine out of ten runway incursions do not. Typically, it is simply a matter of a missed—or misunderstood—clearance, a distraction, or unfamiliarity with an airport or ATC environment. But regardless of the cause, one solution is as simple as it is in our automobiles—just stop and ask. But there are a great many nuances to operating safely in the complex environment of our major air terminals, and that is why this book was written.

The purpose of this volume of the "Controlling Pilot Error" series is to tackle the serious challenge of reducing—or perhaps even eliminating—runway incursions head on, with a focus on individual preparation and personal accountability. Bill Clarke leads us on this journey towards excellence and is more than qualified to do so.

Bill is a graduate of the University of the State of New York and has authored several technical aviation books, including the *Illustrated Guide to Used Airplanes, The Pilot's Guide to Affordable Classics,* and *Aviator's Guide to GPS,* to name just a few. He has also researched and written extensively about radio communications, computer technology, and a wide variety of aviation safety issues. His research into the issue of runway incursions is the most complete you will find, and this manuscript is written with error prevention and mitigation as the central theme. In short, if you don't want to fall victim to an inadvertent runway incursion, you have selected the right book and the right author.

The problem of runway incursions has reached epidemic proportions and has been specifically targeted by national and international aviation safety organizations for elimination. But the bottom line to accomplishing this worthy goal lies, as always, with the pilot in command of any given aircraft. Until we can reach each individual pilot with the skills, knowledge, and *attitude* to operate safely on the ground as well as in the air, we will continue to suffer runway incursion mishaps. Let the battle be joined.

Tony Kern

Runway Incursions

Part 1
Runway Incursions and Airport Operations

1

Runway Incursions Defined

The Federal Aviation Administration (FAA) defines a *runway incursion* as "any occurrence at an airport involving an aircraft, vehicle, person, or object on the ground that creates a collision hazard or results in loss of separation with an aircraft taking off, or intending to take off, landing, or intending to land." The collision hazard or loss of separation can be as simple as an aircraft failing to turn onto a specified taxiway or as serious as an airplane cleared for take off while part of that specific runway is being used for taxi operations. Incursions also can involve airport vehicles, loose or wild animals, or even unforseen police chases.

With this definition and comments in mind, the number of incidents is increasing each year. There were 325 reported runway incursions in 1998, nearly one each day. The problem is getting worse; in 2000, there were 431 runway incursions compared with 321 in 1999, per FAA figures. Incidents in 2001 are running ahead of 2000's pace, with 130 incursions during the first 4 months of

2001 compared with 118 during the same period in 2000. Note that although the FAA officially recognizes (counts statistically) only those runway incursions occurring at controlled airports (towered airports), they can and do occur at all airports.

The FAA reports that during the period of 1997 through 2000, there were 266 million airport operations (at towered airports). There were a total of 1369 runway incursions among those operations, of which three resulted in accidents.

When considered by the numbers, a few hundred reported runway incursions among nearly 60 million operations (arrivals and departures) annually may not seem very large. However, when considered individually, based on the results of specific incursions, the problem can be very serious or very minor—depending on the amount of separation lost between the involved aircraft. For example:

- In November 1999, a United Airlines passenger aircraft taking off from Los Angeles International Airport (LAX) passed over an Aeromexico aircraft that was crossing the runway, missing the latter by less than 100 ft.
- In June 2000, a US Airways Shuttle aircraft landing at LaGuardia Airport (LGA) missed colliding with a corporate turboprop by less than 300 ft.
- In May 2001, the pilots of an American Airlines flight taking off from Dallas Fort Worth (DFW) reported they had to lift off before reaching the proper takeoff speed because a cargo plane was in their path on the runway. Distance between the aircraft was reported to be as small as 10 ft.

The most serious runway incursion to date was at Tenerife in the Canary Islands, where on March 27, 1977, two Boeing 747 aircraft collided on the runway. Tragically, during poor visibility, one aircraft missed a taxiway turnout, and the other began the takeoff roll without permission. They struck head-on, killing 583 persons. This single runway incursion resulted in the worst air disaster in history.

At the Nation's Capital

On May 14, 2001, a private twin-engine plane and a US Airways jetliner bound for West Palm Beach, Florida, nearly collided at the intersection of two runways at Washington's Reagan National Airport (DCA). According to an FAA investigation, a tower controller cleared a Piper twin to land on runway 4 and then mistakenly gave the US Airways Boeing 737 clearance to take off on runway 1, which intersects with runway 4 less than 1000 ft from the Piper's touchdown point (FIG. 1-1).

The controller caught the error 19 seconds later and instructed the Piper pilot to abort the landing. However, the controller was using an incorrect call sign for the Piper; therefore, the pilot did not realize the controller was telling the Piper to abort the landing. The Piper pilot saw the US Airways plane beginning its takeoff roll and braked hard, stopping a mere 200 ft short of the intersection as the jet roared past.

This was classified as a very serious incident of runway incursion by the FAA, and it appears to have been caused by the controller's error.

Necessary Terminology

The following terminology, specific to the problem of runway incursions, will help you in understanding the problem:

Pilot deviation. An action (or inaction) of a pilot that violates any Federal Aviation Regulation (FAR). The

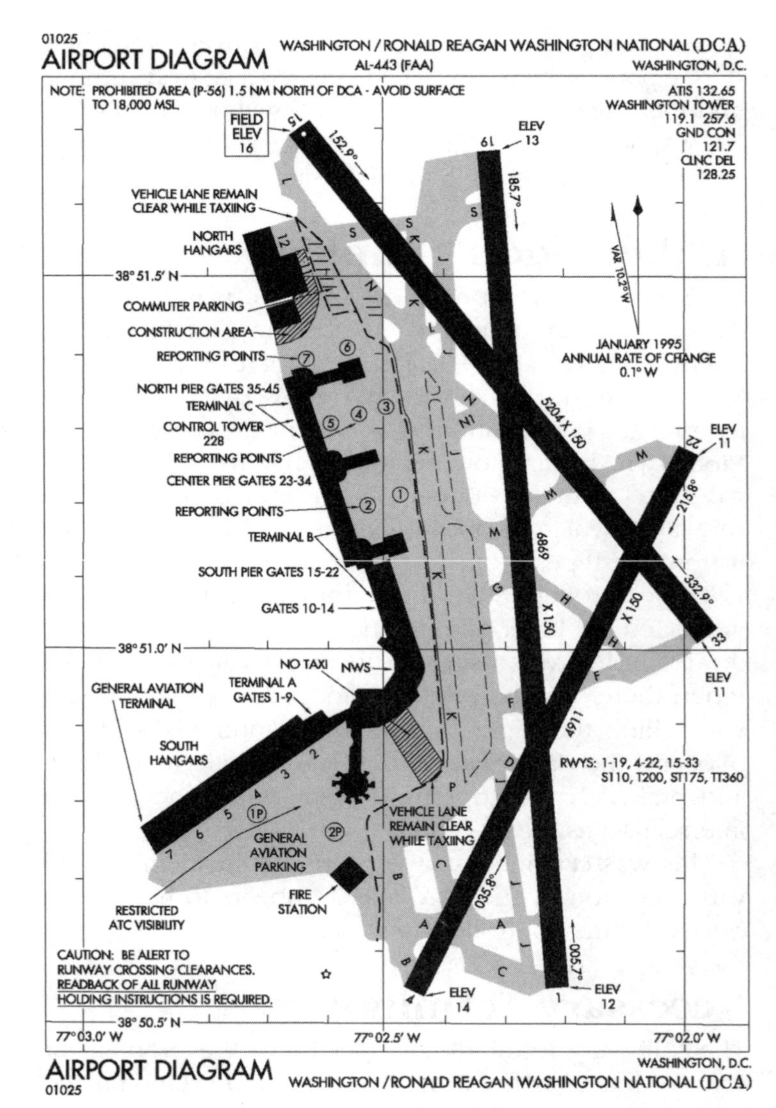

1-1 *Washington/Ronald Reagan National, DCA (this airport diagram is not suitable for navigational purposes).*

most prevalent examples of pilot deviation occur when a pilot fails to obey air traffic control (ATC) instructions to hold short of an active runway (do not cross the runway) or when a pilot selects and uses a specific taxiway without proper clearance.

Operational error. This is an action of an air traffic controller that results in a loss of separation, i.e., less than the FAR-required minimum separation between two or more aircraft or between an aircraft and obstacles (vehicles, equipment, personnel) on runways.

Vehicle/pedestrian deviation. This is the entry or movement on the runway or taxiway area by a vehicle or pedestrian that has not been authorized by ATC. As a side note, this includes taxiing aircraft operated by nonpilots.

Surface incident. This occurs when an unauthorized or unapproved movement (aircraft, vehicle, pedestrian, animal) takes place within the movement area or there is an occurrence in the movement area associated with the operation of an aircraft that affects or could affect the safety of flight. Surface incidents result from pilot deviation, operational error, or vehicle/pedestrian deviation.

Loss of separation. This is an occurrence or operation that results in less than the FAR-prescribed separation between aircraft, vehicles, or objects.

Collision hazard. This is any condition, event, or circumstance that could cause a collision or near-collision and causes the flight crew to take evasive action.

The preceding terminology is the basis for every runway incursion description. Next are the severity catgories, the remaining descriptors used routinely for describing runway incursion events.

Incursion Severity Categories

In 2001, the FAA instituted a system for categorizing runway incursions based on their severity. There are five categories:

Category D. Little or no chance of a collision but meets the definition of a runway incursion. *Example:* An aircraft passed a holding position marking and taxied onto an active runway, but there were no other aircraft operating in the airport's vicinity at the time. There was no need for evasive or corrective action on anyone's part.

Category C. Separation decreases, but there is ample time and distance to avoid a potential collision. *Example:* An aircraft passed a holding position marking and taxied across the far end of an active runway, and there was an aircraft landing on the other end of the runway at the same time. Some evasive or corrective action may have been required, such as the landing aircraft braking harder than normal.

Category B. Separation decreases, and there is a significant potential for collision. *Example:* An aircraft passed a holding position marking and taxied across the near end of an active runway, and there was an aircraft cleared to land on short final. There would be an essential need for evasive or corrective action, such as a go-around on the part of the landing aircraft.

Category A. Separation decreases, and participants take extreme action to narrowly avoid collision.

Example: An aircraft passed a holding position marking and taxied onto and stopped on the near end of an active runway, and there was an aircraft about to touch down on the runway. There would be a critical requirement for a radical evasive action to avoid collision, such as the application of full power, emergency pull-up, and a possible directional correction.

Accident. An incursion that resulted in a runway collision. *Example:* An aircraft passed a holding position marking and taxied onto and stopped on the near end of an active runway, and a landing aircraft collided with the incurring aircraft.

The severity categories, coupled with the previously seen event-description terminology, allow nearly any runway incursion basically to be described in a single sentence. Next, the FAA-developed statistical facts about runway incursion are introduced.

Statistical Facts

The following statistical information was developed by the FAA and serves to indicate where runway incursion problems are stemming from and who is causing them:

- Weather was not a factor in 89 percent of all reported runway incursions.
- Pilots taxiing onto runways or taxiways without clearance accounted for 62 percent of the incursions.
- Low-time pilots (less than 100 hours) accounted for 32 percent of the incursions.
- Pilots landing or departing without clearance accounted for 23 percent of the incursions.

- Pilots unfamiliar with ATC procedures or language accounted for 22 percent of the incursions.
- Pilots unfamiliar with the airport accounted for 19 percent of the incursions.
- Pilot distractions accounted for 17 percent of the incursions.
- Pilots were disorientated or lost in 12 percent of the incursions.
- Pilots landing on the wrong runway accounted for 10 percent of the incursions.
- High-time pilots (greater than 3000 hours) accounted for 10 percent of the incursions.
- Single-engine general aviation (GA) aircraft were involved in 56 percent of the incursions (with the Cessna 172 being the most often reported).

Although ATC and air-carrier pilots do shoulder some of the responsibility, a whopping 69 percent of all runway incursions involved GA aircraft (of all types). However, to keep these statistics in perspective, keep in mind that the GA fleet accounts for 90 percent of the entire U.S. fleet of registered aircraft. Further, GA-type aircraft means anything that is not classified as an air carrier, e.g., corporate aircraft, air taxi aircraft, and small personally owned aircraft.

As a side note, in Europe, where commercial traffic operates at a similar level as in the United States, there are far fewer incidents of runway incursion, the notable difference being a greatly reduced level of GA flying.

Where This Is Going

The FAA has determined by various studies and statistics that the 32 busiest U.S. airports have twice the average number of reported runway incursions. This, no doubt, is due to the large volume of air traffic handled at these locations. Therefore, there is a major concern that proposed changes in ATC systems and procedures—to increase traffic at the busiest U.S. airports—will increase the number of runway incursions.

An interesting statement from the FAA indicated that it did not know why the number of runway incursions is up but suspected that part of the reason was that pilots and controllers were reporting more minor incidents. On the surface, this may seem plausible, because there is a lot of interest in runway incursions at this time. However, could this statement mean that prior to this great interest, runway incursions were overlooked? Further fuel to the fires of FAA statistics is a report from the *Washington Post* that some near-collisions have not been reported as required.

Four-Year Accident Overview

In the years from 1997 through 2000, there were three runway collisions. They include (as provided in a FAA Runway Safety Report):

1. La Guardia Airport (LGA): Operational error involving a privately owned twin-engine aircraft and an airport maintenance vehicle (1997).
2. Sarasota-Bradenton Airport (SRQ): Operational error involving two small privately owned propeller aircraft (2000).
3. Fort Lauderdale (FLL): Vehicle/pedestrian deviation involving an airport truck and a commercial passenger jet (2000).

A Prevention Overview

Over the years, the number of runway incursions has been increasing. Although there appear to be a few high-technology answers to the problem, the primary means of reversing the increasing trend of incursions is extra care and vigilance by pilots.

The three identified areas where pilots can help in the reduction of the number of runway incursions are

1. *Improved communications.* This involves direct communications with ATC and monitoring of communications involving other aircraft. The latter provides an overview of current airport traffic and conditions. The use of consistent terminology is recommended for all involved (pilots and controllers).
2. *Increasing airport knowledge.* This involves using appropriate airport diagrams in preparation for and during ground operations. Knowing where you are, where you are going, and how you will get there—before contacting ATC—can greatly reduce controller/pilot misunderstandings.
3. *Proper cockpit procedures.* This involves maintaining a sterile cockpit, thus allowing full attention to the task at hand and being aware of your surroundings. Reducing outside influences such as unnecessary conversation (explain to your passengers that you are busy, not rude) is a first step.

These three areas of improvement are the responsibility of the pilot or aircraft commander. Hence pilot responsibility is discussed next, via FAA rules and regulations contained in the *Aeronautical Information Manual* (AIM).

Pilot Responsibility

As pilots, we are responsible for our actions. As excerpted from the *Aeronautical Information Manual* (AIM), "The pilot in command of an aircraft is directly responsible for and is the final authority as to the operation of that aircraft."

From the Federal Aviation Regulations (FARs):

Section 91.3. Responsibility and authority of the pilot in command.

(a) The pilot in command of an aircraft is directly responsible for, and is the final authority as to, the operation of that aircraft.

(b) In an in-flight emergency requiring immediate action, the pilot in command may deviate from any rule of this part to the extent required to meet that emergency.

(c) Each pilot in command who deviates from a rule under paragraph (b) of this section shall, upon the request of the Administrator, send a written report of that deviation to the Administrator.

Section 91.123. Compliance with ATC clearances and instructions.

(a) When an ATC clearance has been obtained, no pilot in command may deviate from that clearance unless an amended clearance is obtained, an emergency exists, or the deviation is in response to a traffic alert and collision avoidance system resolution advisory. However, except in Class A airspace, a pilot may cancel an IFR flight plan if the operation is being conducted in VFR weather conditions. When a pilot is uncertain of an ATC clearance, that pilot shall immediately request clarification from ATC.

(b) Except in an emergency, no person may operate an aircraft contrary to an ATC instruction in an area in which air traffic control is exercised.

(c) Each pilot in command who, in an emergency, or in response to a traffic alert and collision avoidance system resolution advisory, deviates from an ATC clearance or instruction shall notify ATC of that deviation as soon as possible.

(d) Each pilot in command who (though not deviating from a rule of this subpart) is given priority by ATC in an emergency, shall submit a detailed report of that emergency within 48 hours to the manager of that ATC facility, if requested by ATC.

(e) Unless otherwise authorized by ATC, no person operating an aircraft may operate that aircraft according to any clearance or instruction that has been issued to the pilot of another aircraft for radar air traffic control purposes.

These rules must be followed by pilots and flight crews. However, pilots are human, and that area—the human part of the problem—is explored next.

The Human Part of the Problem

Although relatively few runway incursions occur when compared with the massive amount of traffic that moves safely through our airports every day, incursions do present a special problem. Each incursion has the potential to put many lives at risk due to the number and proximity of aircraft operating at a particular airport.

Runway incursions occur in a complex and dynamic environment where basic individual causes are difficult to isolate. At the simplest level, runway incursions occur primarily due to human error—an easy place to lay blame and an equally difficult place to fix quickly.

Recognizing that humans are superbly skilled at making decisions under a wide range of circumstances, we must also realize that for a variety of reasons—including forgetting, overlooking, misunderstanding, ignoring, becoming bored or complacent, failing to perceive, having physical limitations, or becoming mentally overloaded (causing disorientation or confusion)—they are not infallible and are prone to error.

Everyday Similarity

Runway incursions are similar in nature to motorists failing to observe proper and lawful automobile traffic controls such as speed limits, stop signs and traffic lights, one-way signs, etc. Although the vast majority of the driving public operates in a safe and proper manner, a small percentage fails (by choice, ignorance, or inattention) to observe the rules and causes accidents. To make more of a parallel between highway driving and flying, there even have been reported incidents of runway incursions that have all the indicators of "road rage," including physical altercations between the participants.

Current Prevention Methods

In-depth study, education, and strict enforcement are the tools generally used to correct motorist problems. Similar tools are being used to reduce the ever-growing number of runway incursions.

NASA's Aviation Safety Reporting System (ASRS) affords the study material for analyzing the problem of runway incursions via its collection of reports about runway incursions. Examples of education about runway incursions include that from the Air Safety Foundation, an arm of the Aircraft Owners and Pilots Association (AOPA), and various FAA programs put on in many locations around the country. Of course, the enforcement of the FARs is handled by the FAA and generally is considered to be the last step—with pilot education and cooperation being preferred.

This overall method of handling a problem is sometimes called *study, educate, enforce* (SEE) and has been successful in many areas, not just flying.

For All of Us

Runway incursions are not caused by pilots alone, or by air traffic controllers, or by vehicle operators, or even by pedestrians. Runway incursions are a problem that all of us in the aviation community share—and must solve.

2

Everyday Airport Operations

There are two kinds of airports—those with a control tower and those without. In other words, towered and nontowered, controlled or uncontrolled. Just to muddy up the waters a little, when a part-time tower (operational only during specified time periods) is closed, the airport is considered nontowered.

Because of Air Traffic Control (ATC) clearance requirements, the procedures used at towered (controlled) airports differ from those used at uncontrolled airports. A pilot approaching an uncontrolled (nontowered) airport generally will enter the downwind leg of the traffic pattern at a 45-degree angle and continue to fly a standard traffic pattern. The same pilot approaching a controlled airport can expect ATC to work him or her into the traffic at nearly any point (base, straight in, etc.) based on the direction of travel, the particular runway(s) in use at the time, and current air traffic load.

This chapter will briefly discuss operations at controlled (towered) airports with runway incursion avoidance in mind. The chapter will look at the roles of the folks involved, check a few rules, and talk a little about how signage and the runway/taxiway lighting systems work. Then the chapter will explore a little about land and hold-short operations (LAHSO) and the problem between pilots and controllers regarding read backs of hold-short clearances. Lastly, the chapter will look at what to do if you lose the capability to communicate with the tower.

Pilot/Controller Roles and Rules

The aircraft pilot and the ATC controller have specific roles when it comes to air traffic and specific rules they must follow while in these roles. Basic information regarding these roles and rules is outlined in the *Aeronautical Information Manual* (AIM), Section 5, Pilot/Controller Roles and Responsibilities:

5-5-1. General

(a) The roles and responsibilities of the pilot and controller for effective participation in the ATC system are contained in several documents. Pilot responsibilities are in the CFRs, and the air traffic controller's are in the FAA Order 7110.65, Air Traffic Control, and supplemental FAA directives. Additional and supplemental information for pilots can be found in the current Aeronautical Information Manual (AIM), Notices to Airmen (NOTAMs), Advisory Circulars (ACs), and aeronautical charts. Since there are many other excellent publications produced by nongovernment organizations, as well as other government organizations, with various updating cycles, questions concerning the latest or most current material can be resolved by cross-checking with the above-mentioned documents.

(b) The pilot in command of an aircraft is directly responsible for and is the final authority as to the safe operation of that aircraft. In an emergency requiring immediate action, the pilot in command may deviate from any rule in the General Subpart A and Flight Rules Subpart B in accordance with 14 CFR Section 91.3.

(c) The air traffic controller is responsible to give first priority to the separation of aircraft and to the issuance of radar safety alerts, second priority to other services that are required but do not involve separation of aircraft, and third priority to additional services to the extent possible.

(d) In order to maintain a safe and efficient air traffic system, it is necessary that each party fulfill their responsibilities to the fullest.

(e) The responsibilities of the pilot and the controller intentionally overlap in many areas, providing a degree of redundancy. Should one or the other fail in any manner, this overlapping responsibility is expected to compensate, in many cases, for failures that may affect safety.

(f) The following, while not intended to be all inclusive, is a brief listing of pilot and controller responsibilities for some commonly used procedures or phases of flight. More detailed explanations are contained in other portions of this publication, the appropriate CFRs, ACs, and similar publications. The information provided is an overview of the principles involved and is not meant as an interpretation of the rules nor is it intended to extend or diminish responsibilities.

5-5-2. Air Traffic Clearance

(a) Pilot

(1) Acknowledges receipt and understanding of an ATC clearance.

(2) Reads back any hold short of runway instructions issued by ATC.

(3) Requests clarification or amendment, as appropriate, anytime a clearance is not fully understood or considered unacceptable from a safety standpoint.

(4) Promptly complies with an air traffic clearance on receipt except as necessary to cope with an emergency. Advises ATC as soon as possible and obtains an amended clearance if deviation is necessary.

NOTE: A clearance to land means that appropriate separation on the landing runway will be ensured. A landing clearance does not relieve the pilot from compliance with any previously issued altitude crossing restriction.

(b) Controller

(1) Issues appropriate clearances for the operation to be conducted, or being conducted, in accordance with established criteria.

(2) Assigns altitudes in IFR clearances that are at or above the minimum IFR altitudes in controlled airspace.

(3) Ensures acknowledgment by the pilot for issued information, clearances, or instructions.

(4) Ensures that read-backs by the pilot of altitude, heading, or other items are correct. If incorrect, distorted, or incomplete, makes corrections as appropriate.

It is the interaction between the pilot(s) and controller(s), while fulfilling these roles and responsibilities that allows air traffic to flow in a timely and safe manner. Next, visual aids are explored, in regard to these roles and responsibilities.

Visual Traffic Aids

While driving to the airport in an automobile, typically you will encounter a number of visual traffic aids (street signs). These visual aids include mileage markers, direction-of-travel indicators, street names, warnings (sharp curve, rough road, deer crossing, etc.), passing and no-passing line markings, lane markings, stop lines, traffic control signals (stop lights), where-to-turn information, and on and on.

The pavement markings, signs, and lighting found at an airport provide similar directional and control information. The theory of runway/taxiway signage, lighting, and marking is no different from that used for the highway: Simply move the traffic through in an expeditious and safe manner, causing as little confusion as possible.

Federal Aviation Regulations (FARs) require that there be consistency and uniformity of airport markings and signs from one airport to another because consistency improves efficiency and uniformity enhances safety through ease of comprehension.

The idea of consistency and uniformity from the FAA is not very different from the way highway signs and markings are similar from state to state.

Airport Signage

There are six types of signs that can be found at airports, although most likely you will not encounter them all at any one given time. The more complex the airport, the more important the signs become to pilots—they are the street signs of the runways and taxiways.

1. *Mandatory instruction signs.* These impart required instructions to the pilot. They have a red background with a white inscription and denote an entrance to a runway, a holding position, a critical area, or a prohibited area. They are rectangular in shape, and they may have arrows indicating direction. Runway holding position signs, runway approach area holding position signs, ILS critical area holding position signs, and no entry signs are all examples of mandatory instruction signs.
2. *Location signs.* These are used to identify the runway/taxiway on which an aircraft is located or a boundary. Runway/taxiway identification signs have a black background with a yellow inscription and a yellow border and are used to identify a taxiway or runway location, the boundary of a runway, or an instrument landing system (ILS)–critical area. They are rectangular in shape, they may be colocated with mandatory instruction signs, and they do not have arrows indicating direction. *Runway boundary signs* have a yellow background with a black inscription consisting of two solid horizontal lines and two dashed horizontal lines (solid lines are placed over the dashed lines). These signs are intended to provide pilots with a visual cue for use in determining when the aircraft is clear of the runway. *ILS-critical-area boundary signs* have a yellow background with a black inscription consisting of two solid horizontal lines connected by solid vertical lines

(resembling a railroad track). These signs are intended to provide pilots with a visual cue for use in determining when the aircraft is clear of an ILS-critical area.

3. *Direction signs.* These identify the designations of intersecting taxiways leading out of an intersection. They have a yellow background with a black inscription. Arrows are used to indicate the direction to the indicated taxiway. Direction signs may be colocated with location signs.
4. *Destination signs.* These indicate a specific destination on the airport. They have a yellow background with a black inscription and always have an arrow indicating the direction to the specified destination. They are used to provide an indication of where runways, aprons, terminals, military areas, cargo areas, fixed based operations (FBOs), and civil aviation areas are located. Easy-to-understand abbreviations generally are used on these signs.
5. *Information signs.* These have a yellow background with black inscription and are used to provide a pilot with information about applicable radio frequencies, specific areas, and noise abatement procedures. They are a locally controlled sign and may vary in size and placement.
6. *Runway distance-remaining signs.* These are placed along the sides of the runway. They have a black background with white numbers indicating the distance of the remaining runway in thousands of feet.

Signs are posted along the sides of runways and taxiways, but they are not the only visual aids encountered at the airport. Next, airport surface markings are described and explained.

Runway/Taxiway Surface Markings

The assorted painted codes on the runways and taxiways are intended to be as easy to understand as possible. To be easy to understand, a simple color code system is used:

1. White markings are used on runways—for runway numbers, the runway centerline, runway side-strip markings (if used), threshold markings, and aiming point. Note, however, that there is a single exception to this rule: Runway shoulder stripes that are used to supplement runway side stripes are yellow.
2. Yellow markings are used on taxiways.

All taxiways have a yellow centerline marking and runway holding position markings whenever they intersect a runway. Taxiway edge markings are used only when there is a need to separate a taxiway from a paved area not intended for aircraft use or to mark the edge of the taxiway (adjacent to a parking ramp). Additionally, taxiways may have shoulder markings and holding position markings for instrument landing system (ILS)–critical areas and for taxiway intersections.

Surface-painted taxiway direction signs. These have a yellow background with a black inscription and are used to supplement direction signs (sometimes used in lieu of a sign where circumstances prevent sign installation). These markings are located adjacent to the centerline on the side indicating the direction of the turn.

Surface-painted location signs. These have a black background with a yellow inscription and are used to supplement location signs. They are to the right of the centerline.

Geographic position markings. These are placed at points along low-visibility taxi routes designated in the airport's surface movement guidance control system (SMGCS) plan and are used to identify the location of taxiing aircraft during low-visibility operations. They are to the left of the centerline and consist of a black outer ring, white ring, and pink center with a letter or number corresponding to a consecutive position along the route (black and white rings may be reversed for visibility purposes on dark surfaces).

Runway holding position markings. These indicate where an aircraft is required to stop and wait for further ATC clearance. They consist of four yellow lines—two solid and two dashed—extending the full width of the taxiway or runway. The solid lines are always on the side where the aircraft is to hold. There are three locations where runway holding position markings will be encountered.

1. *On taxiways.* To identify the locations on a taxiway where an aircraft is required to stop (hold) when it does not have clearance to proceed onto the runway. *Note:* An aircraft exiting a runway is not clear of the runway until all parts of the aircraft have crossed the applicable holding position marking.
2. *On runways.* These are only for land and hold-short operations (LAHSO) and taxiing operations. A sign with a white inscription on a red background is placed adjacent to these holding position markings. The holding

position markings are placed on runways prior to the intersection with another runway or other designated point.

3. *On taxiways located in runway approach areas.* This marking is colocated with the runway approach area holding position sign, at airports where it is necessary to hold an aircraft on a taxiway located in the approach or departure area of a runway so that the aircraft does not interfere with the ongoing operations on that runway. .

Holding position markings for ILS-critical areas. These consist of two yellow solid lines spaced 2 ft apart and connected by pairs of solid lines spaced 10 ft apart extending across the width of the taxiway (similar to railroad tracks). A sign with a red background and an inscription in white is installed adjacent to these hold-position markings.

Holding position markings for taxiway intersections. These consist of a single dashed line extending across the width of the taxiway and are placed on taxiways where ATC normally holds aircraft short of a taxiway intersection. *Note:* If there is no such marking present, the pilot should stop the aircraft at a point that provides adequate clearance from an aircraft on the intersecting taxiway.

Surface-painted holding position signs. These have a red background with a white inscription and supplement the signs located at the holding position. They are only used where the width of the holding position on the taxiway is greater than 200 ft and are located to the left side of the taxiway centerline—on the holding side—and before the holding position marking.

Permanently closed runways. These are marked with yellow crosses (X's) at each end and every 1000 ft. All other surface markings are obliterated. Temporary closures vary from NOTAM notification to yellow crosses at each end of the runway.

Visual aids are great—when they are easily seen. Signs can be lighted, and taxi/landing lights normally will illuminate surface markings. How about the runways and taxiways themselves? Next, the visual delineation of runways and taxiways is described and explained.

Runway/Taxiway Lighting

The lighting used for runways and taxiways is designed to ease taxiway and runway navigation (surface navigation). The lighting system is meant to be easy to understand, and as with surface markings and signage, it is color-coded. Runway and taxiway lighting is intended for use during the hours of darkness and/or during periods of reduced visibility.

Runway lighting is designed to make landing and takeoff operations easier during the hours of darkness and during periods of poor visibility. The system is based on color-coded lights.

Runway edge lights. These are white and outline the edges of runways. On instrument runways, yellow-colored lights replace the white lights on the last 2000 ft.

Runway end-identifier lights (REILs). These are red-colored lights for departing aircraft and green-colored lights for landing aircraft.

Runway centerline lights. These are white, except for the last 3000 ft of runway. Between 3000 and 1000 ft remaining, the lights alternate from red to white, and for the last 1000 ft, the lights are all red.

Taxiway lead-off lights. These consist of alternate yellow and green lights and extend from the runway centerline to an exit taxiway. They are used to expedite movement of traffic from the runway.

Land and hold-short lights. These indicate the hold-short point used for LAHSO landings. They consist of a row of pulsating white lights across the runway and are only in use when LAHSO is in effect.

Taxiway lighting is designed for simplicity while navigating on the surface of the airport during the hours of darkness or low visibility. Taxiway lighting is color-coded to reduce confusion.

Taxiway edge lights. These are blue and outline the edges of taxiways.

Taxiway centerline lights. These mark the centerline of the taxiway. They are steady green lights. This style of lighting is also used on taxi paths (runways, aprons, and ramps).

Clearance-bar lights. These are placed at holding positions on taxiways and taxiway-taxiway intersections. They consist of three in-pavement steady yellow lights.

Runway guard lights. These are flashing yellow lights placed in an elevated position or in the pavement at taxiway-runway intersections to make runway-taxiway intersections more conspicuous.

Stop-bar lights. These consist of a row of red lights extending the width of the taxiway and are placed at the runway holding position. After giving the clearance to proceed, ATC will turn the stop-bar lights off, and the runway lead-on lights will be turned on. *Note:* Never cross stop-bar lights when they are illuminated—even if the controller has given clearance to proceed.

LAHSO

Land and hold-short operations (LAHSO) have become very common at the busier airports as a tool for allowing more traffic to be handled at any given time. The following information about LAHSO is reprinted from the AIM:

4-3-11. Pilot Responsibilities When Conducting Land and Hold-Short Operations (LAHSO)

(a) LAHSO is an acronym for land and hold-short operations. These operations include landing and holding short of an intersecting runway, an intersecting taxiway, or some other designated point on a runway other than an intersecting runway or taxiway.

(b) Pilot responsibilities and basic procedures.

(1) LAHSO is an air traffic control procedure that requires pilot participation to balance the needs for increased airport capacity and system efficiency, consistent with safety. This procedure can be done safely provided pilots and controllers are knowledgeable and understand their responsibilities. The following paragraphs outline specific pilot/operator responsibilities when conducting LAHSO.

(2) At controlled airports, air traffic may clear a pilot to land and hold short. Pilots may accept such a clearance provided that the pilot in command determines that the aircraft can land and stop safely within the available landing distance (ALD). ALD data are published in the special notices section of the Airport/Facility Directory (A/FD) and in the U.S. Terminal Procedures publica-

tions. Controllers also will provide ALD data on request. Student pilots or pilots not familiar with LAHSO should not participate in the program.

(3) The pilot in command has the final authority to accept or decline any land and hold-short clearance. The safety and operation of the aircraft remain the responsibility of the pilot. Pilots are expected to decline a LAHSO clearance if they (pilots) determine that it will compromise safety.

(4) To conduct LAHSO, pilots should become familiar with all available information concerning LAHSO at their destination airport. Pilots should have, readily available, the published ALD and runway slope information for all LAHSO runway combinations at each airport of intended landing. Additionally, knowledge about landing performance data permits the pilot to readily determine that the ALD for the assigned runway is sufficient for safe LAHSO. As part of the preflight planning process, pilots should determine if their destination airport has LAHSO. If so, their preflight planning process should include an assessment of which LAHSO combinations would work for them given their aircraft's required landing distance. Good pilot decision making is knowing in advance whether one can accept a LAHSO clearance if offered.

(5) If, for any reason, such as difficulty in discerning the location of a LAHSO intersection, wind conditions, aircraft condition, etc.,

the pilot elects to request to land on the full length of the runway, to land on another runway, or to decline LAHSO, a pilot is expected to inform air traffic promptly, ideally even before the clearance is issued. A LAHSO clearance, once accepted, must be adhered to, just as any other ATC clearance, unless an amended clearance is obtained or an emergency occurs. A LAHSO clearance does not preclude a rejected landing.

(6) A pilot who accepts a LAHSO clearance should land and exit the runway at the first convenient taxiway (unless directed otherwise) before reaching the hold-short point. Otherwise, the pilot must stop and hold at the hold-short point. If a rejected landing becomes necessary after accepting a LAHSO clearance, the pilot should maintain safe separation from other aircraft or vehicles and should notify the controller promptly.

(7) Controllers need a full read-back of all LAHSO clearances. Pilots should read back their LAHSO clearance and include the words, "Hold short of (runway/taxiway/or point)" in their acknowledgment of all LAHSO clearances. In order to reduce frequency congestion, pilots are encouraged to read back the LAHSO clearance without prompting. Do not make the controller have to ask for a read-back!

(c) LAHSO situational awareness.

(1) Situational awareness is vital to the success of LAHSO. Situational awareness starts with having current airport information in the

cockpit, readily accessible to the pilot. (An airport diagram assists pilots in identifying their location on the airport, thus reducing requests for "progressive taxi instructions" from controllers.)

(2) Situational awareness includes effective pilot-controller radio communication. ATC expects pilots to specifically acknowledge and read back all LAHSO clearances as follows:

ATC "(Aircraft ID) cleared to land runway 6 right; hold short of taxiway bravo for crossing traffic (type aircraft).

"Aircraft: "(Aircraft ID), wilco, cleared to land runway 6 right to hold short of taxiway bravo."

ATC: "(Aircraft ID) cross runway 6 right at taxiway bravo; landing aircraft will hold short."

Aircraft: "(Aircraft ID), wilco, cross runway 6 right at bravo; landing traffic (type aircraft) to hold."

(3) For airplanes flown with two crew members, effective intracockpit communication between cockpit crew members is also critical. There have been several instances where the pilot working the radios accepted a LAHSO clearance but then simply forgot to tell the pilot flying the aircraft.

(4) Situational awareness also includes a thorough understanding of the airport markings, signage, and lighting associated with LAHSO. These visual aids consist of a three-part system of yellow hold-short markings, red and white signage, and in certain cases, in-pavement lighting. Visual aids assist the pilot in

determining where to hold short. Pilots are cautioned that not all airports conducting LAHSO have installed any or all of the preceding markings, signage, or lighting.

(5) Pilots should only receive a LAHSO clearance when there is a minimum ceiling of 1000 ft and 3 statute miles of visibility. The intent of having "basic" VFR weather conditions is to allow pilots to maintain visual contact with other aircraft and ground-vehicle operations. Pilots should consider the effects of prevailing in-flight visibility (such as landing into the sun) and how it may affect overall situational awareness. Additionally, surface vehicles and aircraft being taxied by maintenance personnel also may be participating in LAHSO, especially in operations that involve crossing an active runway.

Final LAHSO Authority

Perhaps the most important part of the LAHSO information from AIM Section 4-3-11 for the pilot to remember is, "The pilot in command has the final authority to accept or decline any land and hold-short clearance. The safety and operation of the aircraft remain the responsibility of the pilot. Pilots are expected to decline a LAHSO clearance if they determine it will compromise safety." Simply stated, this means that you are not required to participate in LAHSO. You may decline and ATC cannot legally pressure you into accepting a LAHSO clearance.

Special LAHSO Indicators

LAHSO introduces a few new terms into flying's vocabulary. The following are position-indicator terminology used solely with LAHSO:

Hold-short point. This is a point on the runway beyond which a landing aircraft with a LAHSO clearance is not authorized to proceed.

Hold-short position marking. This is the painted runway holding position marking located at the hold-short point on all LAHSO runways.

Hold-short position signs. These are red and white holding position signs located alongside the hold-short point.

Land and hold-short lights. These are in-pavement pulsing white lights at the LAHSO hold-short point.

NOTAM

Notice to Airman (NOTAM) information is aeronautical information that could affect a pilot's decision to make a flight. It includes such information as airport or primary runway closures, changes in the status of navigational aids, ILSs, radar service availability, and other information essential to planned en route, terminal, or landing operations (AIM Section 5-1-3).

The specific type of NOTAM that could be useful when operating at a controlled airport is NOTAM (L), which includes such information as taxiway closures, personnel and equipment near or crossing runways (work crews), airport rotating beacon outages, and airport lighting aids that do not affect instrument approach criteria, such as a visual approach slope indicator (VASI).

NOTAM (L) information is distributed locally only and is not attached to hourly weather reports. A separate file of NOTAM (L) is maintained at each flight service station (FSS) for facilities in that area only. Information for out-of-area destinations must be requested from the FSS having responsibility for the area in question.

Clearances and Instructions from ATC

Flying to and from towered airports involves various instructions and clearances from ATC. They are intended to maintain aircraft separation. It is an absolute requirement that both the pilot and the controller understand each instruction, clearance, and acknowledgment of the same. Radio communications must be kept short, simple, and clear—thus providing for easy understandability.

A misunderstanding by either the pilot or the controller can have very serious consequences. To ensure understanding, pilots should read back received clearances. *Note:* Controllers are required to get a read-back of hold-short instructions; therefore, such clearances must be read back at all times.

Controllers issuing clearances use specific words and phrases. The object of the specific words and phrases is ease of understanding—through uniformity and consistency. If you do not fully understand a controller's instructions, immediately ask for clarification. The Pilot/Controller Glossary in the AIM is an excellent source for reviewing the terms and phrases (an abridged version of this glossary appears in Appendix A of this book).

Examples of ATC Instructions

The following are examples of typical instructions a pilot can expect to hear from a controller:

Taxi to. Without holding instructions, a clearance to taxi to any point other than an assigned takeoff runway is a clearance to cross all runways that intersect the taxi route. *Note:* This does not include authorization to taxi onto or cross the assigned takeoff runway at any point.

Taxi to runway xx; hold short of (runway or taxiway). Clearance to begin taxiing, but en route you must hold short of another taxiway or a crossing runway as specified by the controller.

Cross runway xx. You are cleared to taxi across the runway intersecting your taxi route and continue.

Hold short of (runway or taxiway). Do not cross the taxiway or runway specified by the controller. If there is a painted hold line, do not cross it. Do not enter the taxiway or runway.

Cleared for immediate takeoff. Clearance to initiate the takeoff without delay. This type of clearance usually means that another aircraft is on final approach. If you are not ready to start your takeoff roll immediately, do not accept this clearance.

Go around. Abort the landing and go around the traffic pattern again.

Land and hold short. See LAHSO operation for further information.

These simple phrases are those heard most commonly from controllers. Next, the controller's job is explored along with the responsibilities a controller assumes.

Requirement for Controllers

Chapter 1 looked at the responsibilities placed on the pilot. The following information is taken from Chapter 3 of the *Airport Traffic Control Handbook* and is provided to give an overview of what a portion of the controller's job consists of and the responsibilities that accompany that job. This information should prove to be enlightening to many pilots, in particular those having never visited a control tower.

3-1-12. Visually Scanning Runways

(a) Local controllers shall visually scan runways to the maximum extent possible.

(b) Ground control shall assist local control in visually scanning runways, especially when runways are in close proximity to other movement areas.

3-6-1. Equipment Usage

(a) ASDE/AMASS shall be operated continuously to augment visual observation of aircraft landing or departing and aircraft or vehicular movements on runways and taxiways or other areas of the movement area.

(b) The operational status of ASDE/AMASS shall be determined during the relief briefing or as soon as possible after assuming responsibility for the associated control position.

3-6-2. Information Usage

(a) ASDE/AMASS derived information may be used to

(1) Formulate clearances and control instructions to aircraft.

(2) Formulate control instructions to vehicles on the movement area.

(3) Position aircraft and vehicles using the movement area.

(4) Determine the exact location of aircraft and vehicles or spatial relationship to other aircraft/vehicles on the movement area.

(5) Monitor compliance with control instructions by aircraft and vehicles on taxiways and runways.

(6) Confirm pilot reported positions.

(7) Provide directional taxi information, as appropriate.

3-6-3. Identification

To identify an observed target on the ASDE/AMASS display, correlate its position with one or more of the following:

(a) Pilot position report.

(b) Controller's visual observation.

(c) An identified target observed on the ASR or BRITE/DBRITE/TDW display.

3-6-4. AMASS Alert Responses

When the system alarms, the controller shall immediately assess the situation visually and as presented on the ASDE/AMASS display and then take appropriate action as follows:

(a) When an arrival aircraft (still airborne, prior to the landing threshold) activates an alarm, the controller shall issue go-around instructions. (*Exception:* Alarms involving known formation flights, as they cross the landing threshold, may be disregarded if all other factors are acceptable.)

(b) For other AMASS alarms, issue instructions/clearances based on good judgment and evaluation of the situation at hand.

3-7-1. Ground Traffic Movement

Issue by radio or directional light signals, specific instructions that approve or disapprove the movement of aircraft, vehicles, equipment, or personnel on the movement area.

(a) Do not issue conditional instructions that depend on the movement of an arrival aircraft on or approaching the runway or a departure aircraft established on a takeoff roll. Do not say, "Taxi into position and hold behind landing traffic" or "Taxi/proceed across runway 36 behind departing/landing Jetstar." The preceding requirements do not preclude issuing instructions to follow an aircraft observed to be operating on the movement area in accordance with an ATC clearance/instruction and in such a manner that the instructions to follow are not ambiguous.

(b) Do not use the word cleared in conjunction with authorization for aircraft to taxi or equipment/vehicle/personnel operations. Use the prefix taxi, proceed, or hold, as appropriate, for aircraft instructions and proceed or hold for equipment/vehicles/personnel.

(c) Intersection departures may be initiated by a controller, or a controller may authorize an intersection departure if a pilot requests. Issue the measured distance from the intersection to the runway end rounded down to the nearest 50 ft to any pilot who requests and to all military aircraft unless use of the intersection is covered in appropriate directives.

NOTE: Exceptions are authorized where specific military aircraft routinely make intersection takeoffs and procedures are defined in appropriate directives. The authority exercising operational control of such aircraft ensures that all pilots are thoroughly familiar with these procedures, including the usable runway length from the applicable intersection.

(d) State the runway intersection when authorizing an aircraft to taxi into position to hold or when clearing an aircraft for take off from an intersection.

3-7-2. Taxi and Ground Movement Operations

Issue, as required or requested, the route for the aircraft/vehicle to follow on the movement area in concise and easy-to-understand terms. When a taxi clearance to a runway is issued to an aircraft, confirm the aircraft has the correct runway assignment.

NOTE:

1. A pilot's read back of taxi instructions with the runway assignment can be considered confirmation of runway assignment.
2. Movement of aircraft or vehicles on nonmovement areas is the responsibility of the pilot, the aircraft operator, or the airport management.
   (a) When authorizing a vehicle to proceed on the movement area or an aircraft to taxi to any point other than an assigned takeoff runway, absence of holding instructions authorizes an aircraft/vehicle to cross all taxiways and runways that intersect the taxi route. If it is the intent to hold the aircraft/vehicle short of any given point along the taxi route, issue the route, if necessary, and then state the holding instructions.

   (b) When authorizing an aircraft to taxi to an assigned takeoff runway and hold-short instructions are not issued, specify the runway preceded by taxi to, and issue taxi instructions if necessary. This authorizes the aircraft to cross all runways/taxiways that the taxi route intersects except the assigned takeoff runway. This does not authorize the aircraft to enter or cross the assigned takeoff runway at any point.

   (c) Specify the runway for departure and any necessary taxi instructions and hold-short restrictions when an aircraft will be required to hold short of a runway along the taxi route.

(d) Request a read back of runway hold-short instructions when it is not received from the pilot/vehicle operator.

(e) Issue progressive taxi/ground movement instructions when

(1) Pilot/operator requests.

(2) The specialist deems it necessary due to traffic or field conditions, e.g., construction or closed taxiways.

(3) As necessary during reduced visibility, especially when the taxi route is not visible from the tower.

(f) Progressive ground movement instructions include step-by-step routing directions.

(g) Instructions to expedite a taxiing aircraft or a moving vehicle.MW

3-7-3. Ground Operations: Wake Turbulence Application

Avoid clearances that require

(a) Heavy jet aircraft to use greater than normal taxiing power.

(b) Small aircraft or helicopters to taxi in close proximity to taxiing or hover-taxi helicopters.

3-7-4. Runway Proximity

Hold a taxiing aircraft or vehicle clear of the runway as follows:

(a) Instruct aircraft or vehicle to hold short of a specific runway.

(b) Instruct aircraft or vehicle to hold at a specified point.

(c) Issue traffic information as necessary.

Now that the controller's responsibilities have been explained, what happens to a controller when the controller makes an error?

Penalties for Controllers

Errors made by controllers are not routinely pushed under the rug, as some would have you to believe. The normal process of corrective action against a controller is to remove the controller from duty for the first offense of operational error and send the controller for retraining. A second operational error within a certain time period can lead to dismissal.

Fly in the Ointment

Although a full read-back of hold-short instructions is required, the ATC controller is under no requirement to correct any mistakes made during the read-back. In other words, the pilot(s) must be sure to have correctly received and understood all hold-short clearances. This situation is caused by the Interpretive Rule.

The Interpretive Rule was published in the *Federal Register* on April 1, 1999 and deals with pilot responsibility for compliance with ATC clearances and instructions. Although much more detailed in its final state, the rule basically says that if a pilot incorrectly hears a clearance—and operates the aircraft in accordance with the incorrect clearance—the pilot is in violation.

The rule refers to FAR Section 91.123; the FAA regulations "require pilots to comply with air traffic control clearances and instructions" and goes on to state that "pilots do not meet this regulatory imperative by offering full and complete read-back or by taking other action that would tend to expose their error and allow for it to be corrected."

If the pilot hears the clearance incorrectly and reads it back to the controller with the error included, and the controller does not correct the clearance, it is considered that the pilot is in violation. However, couched in the wording of the rule is a statement signifying that the pilot's efforts and intentions of making a read-back (even if the controller fails to correct any erroneous information) may be considered when setting the amount of the FAA sanction.

About Read-Backs

The FAA states in the Interpretive Rule that "Read-backs are a redundancy in that they supply a check on the exchange of information transmitted through the actual clearance or instruction. Full and complete read-backs can benefit safety when the overall volume of radio communications is relatively light; however, they can be detrimental during periods of concentrated communications." The FAA requires read-backs in many instances, yet with this rule, it states that they may be in the way of other communications, specifically interfering during times of high-volume communications, which are exactly the times that communication mistakes are most likely to happen. *Note:* The entire text of this very important ruling is found in Appendix B.

Radio Communications Failure

The following information is excerpted from the AIM and serves to explain what the pilot must do in the event of a radio communications failure.

Arriving Aircraft

1. *Receiver inoperative.* If you have reason to believe that your receiver is inoperative, remain

outside or above the class D surface area until the direction and flow of traffic have been determined; then advise the tower of your type aircraft, position, altitude, and intention to land and request that you be controlled with light signals (reference: Traffic Control Light Signals, paragraph 4-62). When you are approximately 3 to 5 mi from the airport, advise the tower of your position, and join the airport traffic pattern. From this point on, watch the tower for light signals. Thereafter, if a complete pattern is made, transmit your position downwind and/or turning base leg.

2. *Transmitter inoperative*. Remain outside or above the class D surface area until the direction and flow of traffic have been determined; then join the airport traffic pattern. Monitor the primary local control frequency as depicted on sectional charts for landing or traffic information, and look for a light signal that may be addressed to your aircraft. During hours of daylight, acknowledge tower transmissions or light signals by rocking your wings. At night, acknowledge by blinking the landing or navigation lights. To acknowledge tower transmissions during daylight hours, hovering helicopters will turn in the direction of the controlling facility and flash the landing light. While in flight, helicopters should show their acknowledgment of receiving a transmission by making shallow banks in opposite directions. At night, helicopters will acknowledge receipt of transmissions by flashing either the landing or the search light.
3. *Transmitter and receiver inoperative*. Remain outside or above the class D surface area until

the direction and flow of traffic have been determined; then join the airport traffic pattern and maintain visual contact with the tower to receive light signals. Acknowledge light signals as noted earlier.

Departing Aircraft

If you experience radio failure prior to leaving the parking area, make every effort to have the equipment repaired. If you are unable to have the malfunction repaired, call the tower by telephone and request authorization to depart without two-way radio communications. If tower authorization is granted, you will be given departure information and requested to monitor the tower frequency or watch for light signals as appropriate. During daylight hours, acknowledge tower transmissions or light signals by moving the ailerons or rudder. At night, acknowledge by blinking the landing or navigation lights. If radio malfunction occurs after departing the parking area, watch the tower for light signals or monitor tower frequency.

Light Signals

The light signals you see coming from the tower will take the place of radio communications—during a landing. Taking off without functioning radios from a controlled airport will, in most cases, not be allowed. However, the final determination for such a takeoff would rest with ATC.

The signal light codes are

Steady green. Cleared to land.

Flashing green. Return for landing (steady green at that time).

Steady red. Yield to other aircraft and continue to circle.

Flashing red. Do not land.

Alternating red and green. Use extreme caution.

Summary

At this point, runway incursions have been defined and explained, controller/pilot roles and responsibilities have been shown, the visual aids that both use in performing their respective duties have been demonstrated, and what a pilot/controller does when parts of the system, specifically radio communications, fail has been illustrated. In Part 2 of the book, examples of what went wrong when errors happened and runway incursions occurred are given with analysis and commentary.

Part 2
Case Studies

99140

LEGEND

INSTRUMENT APPROACH PROCEDURES (CHARTS)

AIRPORT DIAGRAM/AIRPORT SKETCH

Runways

Hard Surface — Other Than Hard Surface — Stopways, Taxiways, Parking Areas, Water Runways — Displaced Threshold

Closed Runway — Closed Taxiway — Under Construction — Metal Surface — Runway Centerline Lights

ARRESTING GEAR: Specific arresting gear systems; e.g., BAK12, MA-1A etc., shown on airport diagrams, not applicable to Civil Pilots. Military Pilots refer to appropriate DOD publications.

uni-directional — bi-directional — Jet Barrier

REFERENCE FEATURES

Buildings..■

Tanks..●

Obstructions...Λ

Airport Beacon #...★

Runway Radar Reflectors..⧗

Control Tower #..■

Helicopter Alighting Areas

Negative Symbols used to identify Copter Procedures landing point....................● ■ ■ ▲ ■

Runway TDZ elevation....................TDZE 123

Runway Slope................................0.3% DOWN / 0.8% UP
(shown when runway slope exceeds 0.3%)

NOTE:
Runway Slope measured to midpoint on runways 8000 feet or longer.

U.S. Navy Optical Landing System (OLS) "OLS" location is shown because of its height of approximately 7 feet and proximity to edge of runway may create an obstruction for some types of aircraft.

Approach light symbols are shown in the Flight Information Handbook.

Airport diagram scales are variable.

True/magnetic North orientation may vary from diagram to diagram

When Control Tower and Rotating Beacon are co-located, Beacon symbol will be used and further identified as TWR.

Runway length depicted is the physical length of the runway (end-to-end, including displaced thresholds if any) but excluding areas designated as stopways. Where a displaced threshold is shown and/or part of the runway is otherwise not available for landing, an annotation is added to indicate the landing length of the runway; e.g., Rwy 13 ldg 5000′.

Coordinate values are shown in 1 or ½ minute increments. They are further broken down into 6 second ticks, within each 1 minute increments.

Positional accuracy within ±600 feet unless otherwise noted on the chart.

NOTE:
All new and revised airport diagrams are shown referenced to the World Geodetic System (WGS) (noted on appropriate diagram), and may not be compatible with local coordinates published in FLIP. (Foreign Only)

Runway Weight Bearing Capacity/or PCN Pavement Classification Number is shown as a codified expression.
Refer to the appropriate Supplement/Directory for applicable codes e.g.,
RWY 14-32 S75, T185, ST175, TT325
PCN 80 F/D/X/U

Runway Slope
FIELD ELEV 174
Rwy 2 ldg 8000′
Displaced Threshold
BAK-12
0.7% UP
Runway Identification
20
2
9000 X 200
023.2°
1000 X 200
Runway End Elevation
ELEV 164
Runway Dimensions (in feet)
Runway Heading (Magnetic)
Stopway Dimensions (in feet)

SCOPE

Airport diagrams are specifically designed to assist in the movement of ground traffic at locations with complex runway/taxiway configurations and provide information for updating Computer Based Navigation Systems (I.E., INS, GPS) aboard aircraft. Airport diagrams are not intended to be used for approach and landing or departure operations. For revisions to Airport Diagrams: Consult FAA Order 7910.4B.

LEGEND

P2-1 *Airport diagram legend.*

The following pages contain case studies of actual runway incursions based on reports from NASA's Aviation Safety Reporting System (ASRS) files. Each case study includes a narrative description of the incident, what preventive preparation(s) was made prior to the incident, a postincident analysis, and the lessons learned from the incident. The narrative section of each incident is as it appeared in the original report—in the words of the reporting person (pilot, controller, or other).

Identifying items are removed by the ASRS system; therefore, you will not see specific flight numbers, exact times, N-numbers, etc., in these reports.

The included airport diagrams are intended to help you to understand the taxi paths, incursions, potential incursions, and other mistakes recorded in this book. They are as taken from the various airport facilities directory products and other NOAAs and public sources.

WARNING: *None* of the included airport diagrams are suitable for, or are intended for, navigational purposes and as such are deemed suitable only for reference use while reading this book.

The legend for airport diagrams can be seen in FIG. P2-1.

Use the AIM

Colored examples of runway/taxiway signage, markings, and lighting can be found in the *Aeronautical Information Manual* (AIM), Chapter 2, Sections 1, 2, and 3.

3

Departure Runway Incursions

Runway incursions during departure operations occur while the aircraft in question is on the ground. This means that the aircraft is moving slowly—taxiing. The most frequent locations for departure incursions are while crossing runways or while taxiing onto an active runway.

In this chapter, the Aviation Safety Reporting System (ASRS) reports examined for the case studies all resulted from incidents at towered (controlled) airports. Some of the incidents involved pilot error, which the Federal Aviation Administration (FAA) refers to as pilot deviation resulting from the violation of any Federal Aviation Regulations (FARs). Other incidents came from operational errors, i.e., those errors attributed to air traffic control (ATC) resulting in the loss of separation. In both types of incursions, those caused by pilot deviations and those caused by operational errors, there were actions that could have been taken to prevent the incident from occurring.

CASE 1

The Big Picture

ASRS accession number: 454710/454815

Month and year: November 1999

Local time of day: 1801 to 2400

Facility: ATL, William B. Hartsfield Atlanta Airport (FIG. 3-1)

Location: Atlanta, GA

Flight conditions: VMC

Aircraft 1: B727

Aircraft 2: L-1011

Pilot of aircraft 1: Captain, 8000 hours; first officer, 9000 hours;` second officer

Pilot of aircraft 2: Captain

Reported by: Pilot of aircraft 1

Incident description: Runway incursion

Incident consequence: Takeoff aborted, and controller issued new takeoff clearance.

NARRATIVE ACR X WAS CLRED FOR TKOF BY ATLANTA TWR (119.5). ATLANTA TWR ALSO USES ANOTHER FREQ FOR RWY 26R FOR LNDG AND XING RWY 26L. WE JUST STARTED OUR ROLL WHEN THE CAPT CALLED 80 KTS AND THE TWR CALLED TO SAY DISCONTINUE THE TKOF BECAUSE OF XING TFC AT THE W END. WE DID STOP, TAXI BACK, CHK OUR CHARTS FOR BRAKE ENERGY, AND TOOK OFF AGAIN.

SUPPLEMENTAL INFO FROM ACN 454815: THE TKOF WAS NORMAL AND I WAS MOSTLY INSIDE CONFIRMING THE ENG INSTS AND PWR SETTINGS. AS I WAS MAKING THE 80 KT CALL TWR SAID SOMETHING, WHICH I DID NOT UNDERSTAND SINCE I WAS TALKING AT THE SAME TIME. TWR SAID ACR X ABORT A SECOND TIME AND WE PULLED THE THROTTLES TO IDLE. I LOOKED TO THE

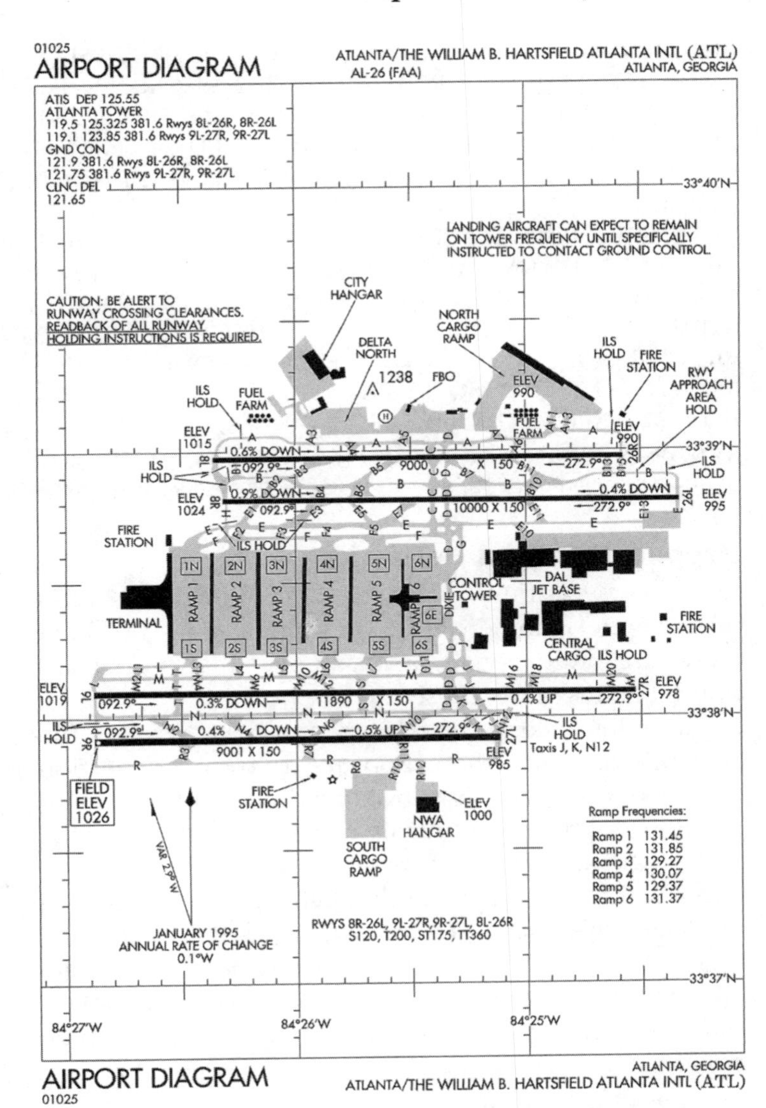

3-1 *Atlanta International Airport, ATL (this airport diagram is not suitable for navigational purposes).*

END OF THE RWY TO SEE AN L1011 FOLLOWED CLOSELY BY A B727 XING THE RWY. THIS SIT WAS CAUSED BY ATLANTA TWR INSISTING ON ALLOWING DIFFERENT ACFT USING THE SAME RWY, BUT BE CTLED ON DIFFERENT RADIO FREQS UNLIKE ANY OTHER ARPT I AM AWARE OF.

ATLANTA ALLOWS ACFT TO TAKE OFF ON 125.32, BUT ACFT XING THE RWY AFTER LNDG ON RWY 26R ON FREQ 119.5. PLTS ARE TAKEN OUT OF THE LOOP. WHY CAN'T ATLANTA DO THIS LIKE DFW AND OTHER ARPTS AND REQUIRE LNDG ACFT HOLD SHORT OF RWY 26L AND MONITOR 125.32. ALSO I ALLOWED TWR TO RUSH ME INTO AN EXPEDITED DEP.

SYNOPSIS ATL LCL CTLR CLRED B727 FOR TKOF, BUT THEN ABORTED THE FLT BECAUSE OF TFC XING THE RWY.

Preparation and Remarks

Atlanta is a very busy airport, the aircraft involved in this incident are commercial air transports, and the flight crews are professionals. It is assumed that proper airport diagrams were in use.

The pilot making the ASRS report was aware of the surroundings, and there appears to be no confusion about taxiway signs or runway marking. Radio communications for the departing aircraft were normal, and the start of the takeoff was normal.

All indications were that the flight crew was doing its work as expected and was properly prepared for the flight. Note, however, that the captain does indicate allowing the tower to rush him and that cockpit conversation caused a radio message from the controller to be missed.

Postincident Analysis

The incident was caused by ATC allowing an aircraft to taxi across an active runway; however, it was brought to a safe conclusion by an alert controller seeing that an aircraft had been mistakenly cleared for takeoff from a runway that was being crossed by an aircraft from another active runway. Thus the controller ordered the aircraft to abort the takeoff.

This incident would be classified as an operational error, which was indicated on the original report under "Problem areas: ATC human performance." The runway incursion severity category of this incident would be category C because the aircraft that caused the aborted takeoff was crossing the far end of the active runway.

Six specific problems are noted from the ASRS report:

Problem 1. The captain indicated being rushed into an expedited departure: "ALSO I ALLOWED TWR TO RUSH ME INTO AN EXPEDITED DEP."

Problem 2. The captain was "inside" checking instruments and power settings: "THE TKOF WAS NORMAL AND I WAS MOSTLY INSIDE CONFIRMING THE ENG INSTS AND PWR SETTINGS."

Problem 3. The captain was talking during a radio transmission, thereby missing the first instruction for abort: "AS I WAS MAKING THE 80 KT CALL TWR SAID SOMETHING, WHICH I DID NOT UNDERSTAND SINCE I WAS TALKING AT THE SAME TIME."

Problem 4. This particular airport, Atlanta (ATL), uses more than a single radio frequency for the control of airplanes on multiple active runways: "ATLANTA ALLOWS ACFT TO TAKE OFF ON 125.32, BUT ACFT XING THE RWY AFTER LNDG ON RWY 26R ON FREQ 119.5."

Problem 5. Pilots and controllers need the "big picture" to do their jobs. The problem of two frequencies being used at an airport for air traffic control is not easily addressed; however, it does lead to confusion. In this incident, multiple frequency usage did lead to active aircraft being totally unaware of each other.

Problem 6. The report indicates that controllers operating the active runways (including taxi to and from) may not always have been aware of each other's actions or aircraft activity.

Lessons Learned

This incident exemplifies how small errors can compound into a large incident.

Lesson 1. As to ATC, no matter how the workload increases, an orderly procedure must be maintained at all times. The first indicator of things about to go awry is the captain allowing the tower to rush the flight into an expedited departure. The rush from the tower shows that

(a) The Atlanta tower was in a hurry to get and keep traffic moving. In such a frenzy, as occurs during certain time periods of high-volume traffic, it is quite possible for orderly control to break down, causing the controllers to make errors—errors that under normal circumstances would not be made.

(b) The pilot should not be sucked into this frenzy by allowing the tower to push or rush any procedure. The result of such rushes can include interruption of normal task patterns with a potential for some details to be overlooked.

Lesson 2. Heads up and look out. Although the captain reports performing cockpit duties during the takeoff, extra eyes observing other airport activity (looking outside the cockpit) are always encouraged.

One must recognize, however, that there are cockpit duties to attend to during takeoff that require attention.

Lesson 3. Maintain a sterile cockpit during takeoff operations. The first "Abort" radio message from the tower was missed due to cockpit conversation. A sterile cockpit environment should be maintained during takeoff operations, meaning no conversation unless absolutely required for the safe operation of the aircraft. This does represent a fine line: Where do safe operation and no talking relate to each other? Certainly there is justification for both, leaving the final decision to the aircraft commander.

Lesson 4. Under normal single-frequency airport operations, all the active aircraft are (or should be) monitoring the same radio traffic from the tower—making them aware of each other. An alternative means of creating the same awareness is to require aircraft to monitor more than a single radio frequency. Of course, the latter brings more confusion into the cockpit.

Lesson 5. Controllers must be aware of all aircraft activity on the airport that may affect or be affected by their actions.

CASE 2

Missed Hold-Short Line

ASRS accession number: 454940

Month and year: November 1999

Local time of day: 0601 to 1200

Facility: PVD, Theodore Francis Green State Airport (FIG. 3-2)

Location: Providence, RI

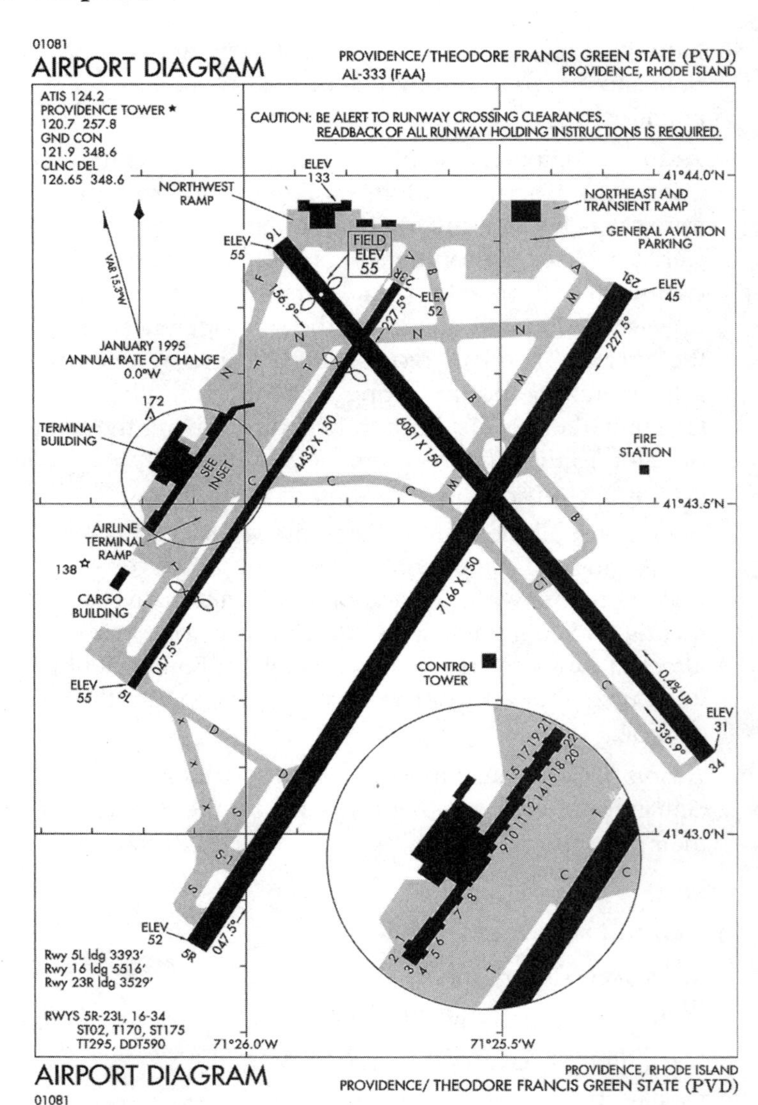

3-2 *Providence/Theodore Francis Green State Airport, PVD (this airport diagram is not suitable for navigational purposes).*

Flight conditions: VMC

Aircraft 1: Saab Scania

Pilot of aircraft 1: Captain, 8500 hours; first officer

Reported by: Pilot of aircraft 1

Incident description: Near runway incursion

Incident consequence: No action taken

NARRATIVE WE WERE INSTRUCTED TO 'TAXI TO RWY 34 VIA TXWY C HOLD SHORT OF RWY 5R.' WHILE TAXIING AND RUNNING CHECKLIST WE APPROACHED THE RWY AND I NOTICED AN OLD HOLD SHORT LINE THAT WAS PAINTED BLACK (IT WAS REFLECTING SUNLIGHT). UPON APPROACHING THE HOLD SHORT LINE GND TOLD US TO 'STOP.' AT THAT TIME I REALIZED THE HOLD-SHORT LINE I WAS APPROACHING WAS NOT THE CORRECT ONE. WE STOPPED SAFELY BEFORE THE RWY AND AN ACFT DEPARTED.

SYNOPSIS SF34 CREW HAD DIFFICULTY IDENTIFYING THE CORRECT HOLD SHORT LINE AT PVD.

Preparation and Remarks

The pilot making the ASRS report was aware of the surroundings, and it is assumed that the correct airport diagram was in use at the time. The aircraft was taxiing, during which time the pilot became confused and unable to identify the correct hold short line. Note the mention in the narrative of "WHILE TAXIING AND RUNNING CHECKLIST..." Note also that the positions of hold-short markings and signs are not included on airport diagrams.

Postincident Analysis

This incident was brought to a safe conclusion when an alert controller told the flight crew to "STOP" the aircraft, thus preventing a runway incursion onto an active runway and possibly into the path of another aircraft.

This incident would be classified as a pilot deviation, as was indicated in the original report under "Problem areas: Flight crew human performance." The runway incursion severity category would be category B because an abrupt stop was necessary to prevent the aircraft from actually entering the active runway for departing aircraft.

Three specific problem areas are noted from this ASRS report:

Problem 1. The captain indicated the observation of an "OLD HOLD SHORT LINE THAT WAS PAINTED BLACK (IT WAS REFLECTING SUNLIGHT)." Under some circumstances, runway markings will appear to be black in color.

Problem 2. The captain became confused about the location of the hold-short line. Observing outside the cockpit would have given visual clues indicating the intersecting point of the taxiway and the runway. A look at the airport diagram shows the taxiway intersecting and crossing runway 5R/23L and ending at runway 34. When nearing the intersecting point with runway 5R/23L, the flight crew should have been able to see the runway—and possibly any aircraft on it waiting to take off (aided by the departing aircraft having landing lights on during takeoff).

Problem 3. The pilot appears to have been less attentive to what was going on outside the cockpit than within. There was a departing aircraft on runway 5R/23L—hence the reason for the hold-short instruction from ATC.

Lessons Learned

This incident was minor in nature because there was no actual runway incursion. However, it does show how important it is to have an alert controller—since it was the controller's radio call to "STOP" that prevented the incursion and possible aftermath. The pilot missed seeing the runway and the departing aircraft on same.

Lesson 1. Under some lighting conditions, hold-short lines and other runway/taxiway markings can appear to be black. Unused markings should be removed in their entirety or obliterated—not just painted over. There is no indication of signs, only pavement markings.

Lesson 2. Be observant and keep heads up and outside the cockpit while taxiing (in this example, the pilot was too much inside the cabin). There are other clues than just the lines painted on the taxiway surface. For example, the runway (5R) intersects taxiway C at right angles. This should have been a very visible clue that this was the hold location and the runway not to cross.

Lesson 3. If you become confused while taxiing, stop the aircraft (not on a runway), and ask the controller for assistance. Taking a few moments of time and getting assistance may prevent a small error from becoming a major incident.

Lesson 4. Hold lines and other surface markings should be supplemented with colocated signs.

CASE 3

The Inexperienced Pilot

ASRS accession number: 455070

Month and year: November 1999

Local time of day: 0601 to 1200

Facility: ATL, William B. Hartsfield Atlanta Airport (FIG. 3-3)

Location: Atlanta, GA

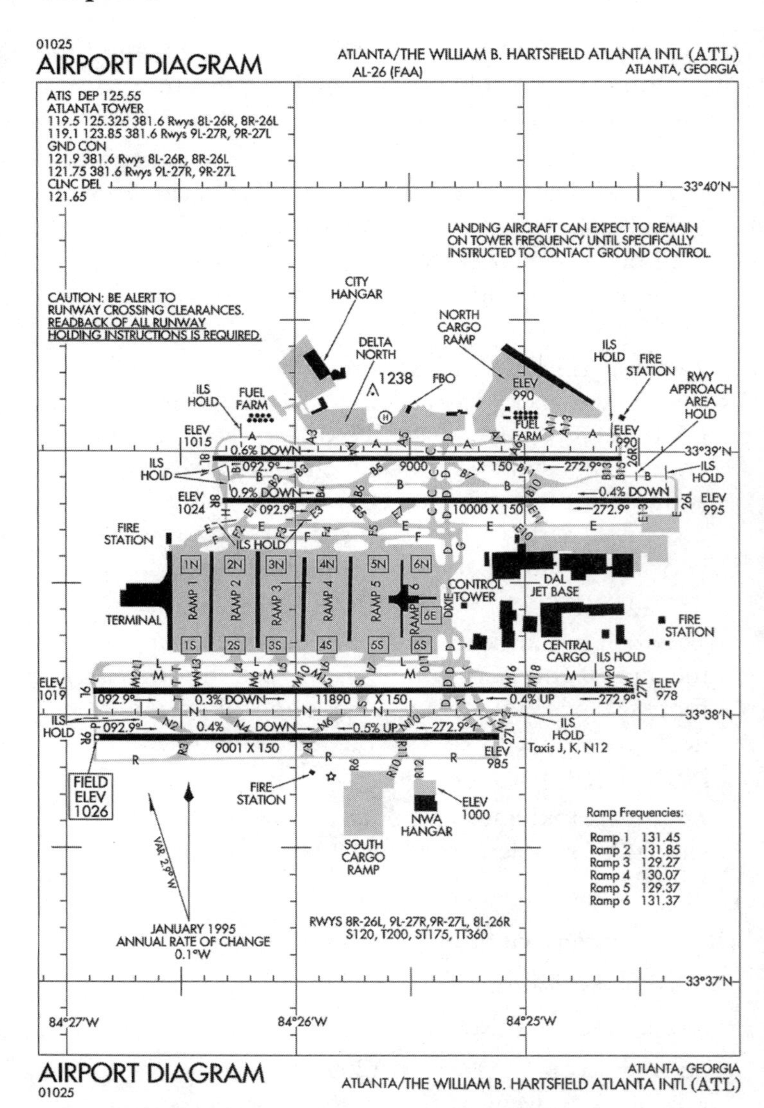

3-3 *Atlanta International Airport, ATL (this airport diagram is not suitable for navigational purposes).*

Flight conditions: VMC

Aircraft 1: SMA (GA small airplane)

Pilot of aircraft 1: Single pilot

Reported by: Controller

Incident description: Runway incursion

Incident consequence: None (detected after the fact)

NARRATIVE ACFT X WAS INSTRUCTED TO HOLD SHORT OF RWY 8L AT TXWY D. ACFT X CROSSED RWY 8L AT TXWY D WITHOUT CLRNC. THIS PLT APPEARED TO ONLY HAVE A PLT'S LICENSE WITH LOW TIME. THESE PLTS SHOULD BE PROHIBITED FROM COMS TO ATL, ORD, LAX, AND ARPTS LIKE THESE. THE LOW TIME PLT DOESN'T POSSESS THE SKILLS TO SAFELY OPERATE AT SOME OF THE WORLD'S BUSIEST ARPTS. IF SOMETHING HAPPENS WITH ONE OF THESE PVT PLTS, YOU CAN COUNT ON SEEING THIS RPT AGAIN.

SYNOPSIS ATCT CTLR AT ATL BELIEVES LOW TIME PLTS SHOULD BE PROHIBITED FROM OPERATING AT LARGE ARPTS.

Preparation and Remarks

The controller making the report appears, based on the report's word choice and demeanor, to be rather upset with the GA aircraft that crossed runway 8L. It is assumed that the controller is familiar with ATL, and from the report, it is assumed that the pilot was not.

Radio communications with the GA aircraft appear to have been normal, and there is no indication of difficulties in comprehension on the part of the GA pilot.

There is no indication of the pilot making a taxi instruction read-back to the controller for the hold-short instruction. Further, there is no indication that the controller asked for a read-back when none was given.

There is no indication that the controller attempted to have the aircraft on taxiway D stop when it became apparent that an incursion was about to happen.

Postincident Analysis

It is apparent from this report that no damage was done by this runway incursion and that there was no separation compromise. The incursion is classified as a pilot deviation—not having done what ATC instructed. In this instance, the pilot did not hold short of the runway. The runway incursion severity category is category D because there was no need for evasive or corrective action on anyone's part.

However, the incident does point out several very significant problem areas:

Problem 1. Operations at busy airports by low-time pilots or otherwise inexperienced pilots can be hazardous. The large amount of radio traffic alone can be overwhelming to an inexperienced pilot at a busy airport. The push and rush of intense air traffic further complicates the situation.

Problem 2. There was a lack of radio communication between the pilot and the controller—as indicated by the lack of a read-back. A major concern about this problem is the controller not asking the GA aircraft's pilot to give a read-back of the hold-short taxi instruction.

Problem 3. An alert controller should have seen the incursion coming and made an attempt at preventing it by warning the pilot or instructing the pilot to stop. Again, the controller did nothing to prevent this incursion from happening.

Problem 4. Why did the controller take no action to prevent the incursion and yet take the time to complete and forward an ASRS report of the incident?

Lessons Learned

Although this incident had a benign result, it could have been much more serious than reported. There was no loss of any separation, and no other aircraft had to make any adjustments due to the incursion.

Lesson 1. Although not proven to be a factor in this incident, there is no doubt that inexperienced pilots should stay away from the very large and busy airports. The complex ATC frequency schemes and large volumes of radio traffic can be very confusing—even to seasoned pilots.

Lesson 2. Communications between the pilot and controller must be complete and accurate. All hold-short taxi instructions are to be read back to the controller [AIM Section 4-3-18(a)8)]: "Pilots should always read back the runway assignment when taxi instructions are received from the controller. The controllers are required to confirm runway hold-short assignment when they issue taxi instructions." The simple procedure of reading back taxi instructions prevents misunderstandings and also serves as a receipt of the instruction by the pilot.

Lesson 3. The pilot ultimately is responsible for operation of the aircraft (FAR Section 91.3).

Lesson 4. As a pilot, do not expect others to correct your mistakes. As in this incident, the controller did nothing to prevent the incursion except issue taxi instructions and make an ASRS report. The pilot is responsible to get it right and to follow it exactly (FAR Section 91.123).

CASE 4

Can't See the Lines

ASRS accession number: 456726

Month and year: December 1999

Local time of day: 1201 to 1800

Facility: SMX, Santa Maria Public—Captain G. Allan

Hancock Field (FIG. 3-4)

Location: Santa Maria, CA

Flight conditions: VMC

Aircraft 1: Tobago TB-10C

Pilot of aircraft 1: Single pilot, 1495 hours

Reported by: Pilot

Incident description: Runway incursion

Incident consequence: FAA reviewed incident with pilot.

NARRATIVE I ARRIVED AT THE RUNUP AREA FOR RWY 30 AT SANTA MARIA'S HANCOCK FIELD. DID MY ACFT'S RUNUP AND HEADED FOR WHAT I THOUGHT WAS THE HOLD SHORT LINE FOR THE RWY. TWR CALLED ME, AFTER CLRING ME FOR TKOF, THAT I HELD SHORT AT THE WRONG LINE, PAST THE HOLD SHORT LINE. THE HOLD-SHORT LINE, AT THIS TIME OF DAY, IS NOT VISIBLE BECAUSE OF THE ANGLE OF THE SUN. THE AFTERNOON SUN CAUSES THE HOLD SHORT LINE TO BLEND IN WITH THE TXWY RUNUP AREA.

SYNOPSIS A TB-10C PLT, TAXIING AT SMX, CROSSED THE HOLD SHORT LINE FOR RWY 30. THE PLT LATER FOUND OUT THAT, AT THE PARTICULAR TIME OF AFTERNOON HE WAS TAXIING, THE HOLD SHORT LINE IS INVISIBLE DUE TO THE ANGLE OF THE SUN.

Preparation and Remarks

The pilot making the report indicated that the preflight preparation was normal and that the taxi clearance was understood. There was no indication of radio communications difficulty. All indications show that until the point of crossing the hold-short line, everything was normal and as expected. It may be that the pilot was not familiar with the Santa Maria airport.

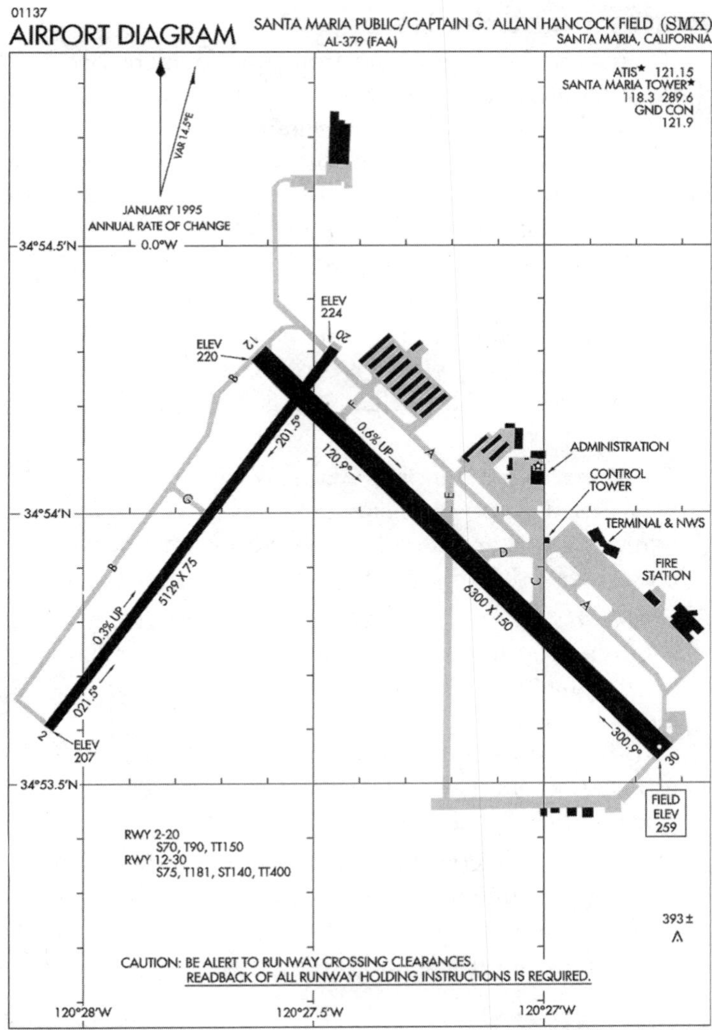

3-4 *Santa Maria Public, SMX (this airport diagram is not suitable for navigational purposes).*

Postincident Analysis

There was no damage done from this incident because there was no indication of separation between aircraft lost. This runway incursion is classified as a pilot deviation—ATC instructions were not followed. The category of this runway incursion would be category D because no other aircraft was involved (or mentioned on the report).

Four specific problems are noted from the ASRS report:

Problem 1. The hold-short line is difficult to see under certain conditions, which is a physical problem that can be addressed by placing signs at the same location.

Problem 2. No mention is made to indicate if there was a holding position sign next to the taxiway markings. It is assumed from the report that none exists (see prior problem).

Problem 3. The general appearance of the run-up area is different from that of a runway. This should have been visually apparent to the pilot. A look at the airport diagram shows the general shape of the taxiway, run-up area, and runway mentioned in this report. Use of an airport diagram should have prevented this incident.

Problem 4. The controller was able to tell the pilot about the incursion when the takeoff clearance was issued. Why not tell the pilot before the incursion and prevent it from happening?

Lessons Learned

This incident shows how naturally occurring conditions sometimes can cause runway incursions. In this specific incident, the pilot was unable to see the hold-short surface markings due to the angle of the sun. Late in the day and in a low November position, there is little doubt that the sun's glare was directly in the pilot's eyes.

Lesson 1. The hold-short line was expected, and the glare of the sun was a known factor. The use of a sun visor and/or proper sunglasses may have helped the pilot by reducing the glare, thus making the painted taxiway markings more visible.

Lesson 2. Look around—a runway is long and straight with its edges disappearing into the distance (in this case to the right of the pilot as runway entry was made). In this case, the runway was away from the sun, making it easy to see. Run-up areas do not visually resemble runways—look for visual clues.

Lesson 3. No indication is made in the ASRS report of the existence of a holding position sign, which normally is placed adjacent to the taxiway surface markings. If such a sign was present, it may have been more easily seen than the surface taxiway markings. Nevertheless, better observation on the part of the pilot is recommended.

Lesson 4. You cannot count on a controller to correct mistakes, because controllers are not always observing the airport surface and its activity. An individual controller, for unknown reasons, may choose not to inform the offending pilot until after the fact.

CASE 5

Clear Communications

ASRS accession number: 456345/456449

Month and year: December 1999

Local time of day: 1201 to 1800

Facility: MDW, Chicago Midway Airport (FIG. 3-5)

Location: Chicago, IL

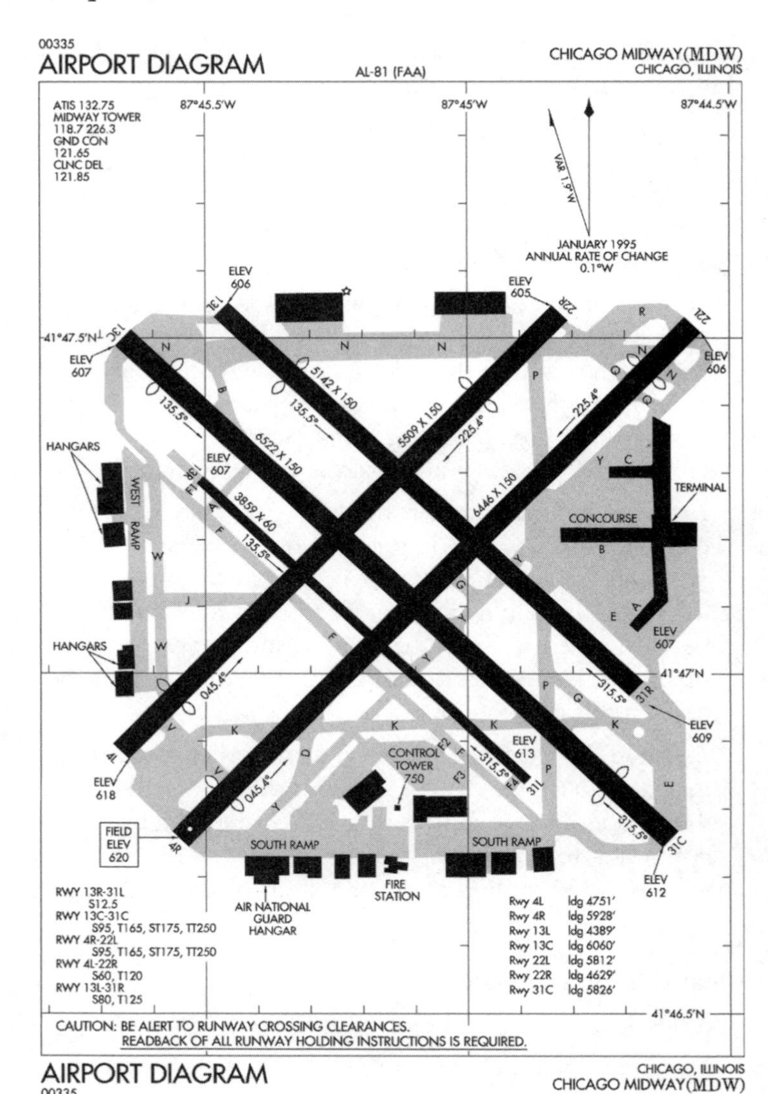

3-5 *Chicago Midway, MDW (this airport diagram is not suitable for navigational purposes).*

Flight conditions: VMC

Aircraft 1: B737

Aircraft 2: Unknown

Pilot of aircraft 1: Captain, 14,900 hours; first officer, 5000 hours

Pilot of aircraft 2: Unknown

Reported by: Pilot(s) of aircraft 1

Incident description: Runway incursion

Incident consequence: Controller issued alert

NARRATIVE. WHILE HOLDING SHORT OF RWY 13C IN MDW, WE (FLT 5XY) HEARD TWR CLR FLT 3XY INTO POS AND HOLD ON RWY 13C. WE ASSUMED THE CALL SIGN TO BE IN ERROR AND THE FO READ BACK THE CLRNC WITH EMPHASIS ON THE 5XY TO CORRECT THE CALL SIGN. TWR DID NOT RESPOND AND WE HAD NO INDICATION WE WERE BLOCKED, SO WE PROCEEDED ON TO RWY 13C WITH A SLIGHT BACK TAXI TO GET FULL LENGTH. DURING THIS SOMEONE WAS CLRED FOR TKOF. THE FO AND I WERE CONFUSED BY THE CLRNC BUT NOTICED LNDG LIGHTS AT THE OTHER END OF THE RWY WHEN WE GOT ALIGNED, SO WE BEGAN TO EXIT THE RWY WHEN TWR CALLED FOR THE ACFT ON THE TXWY (US) TO EXIT. CONTRIBUTING FACTORS INCLUDE: 1) SIMILAR CALL SIGNS THAT WE WERE NOT AWARE OF 2) SIMILAR SOUNDING RWY, IE, RWY 13C AND RWY 31C 3) OPPOSITE DIRECTION DEP WITHOUT NOTICE TO US 4) ASSUMING THAT WE WOULD BE NEXT FOR RELEASE AND NOT GETTING POSITIVE CONFIRMATION OF THE CLRNC.

SUPPLEMENTAL INFO FROM ACN 456449: ASSUMING THAT THE XMISSION WAS MEANT FOR US, WE ANSWERED INQUISITIVELY 'ROGER, THAT'S ACR FLT 5XY, POS AND HOLD RWY 13C,' AND CAUTIOUSLY PROCEEDED ONTO RWY 13C. ACFT FLT 3XY CONTINUED TKOF ROLL AND DEPARTED NORMALLY.

CONTRIBUTING FACTORS: THE USE OF RWY 31C FOR DEPS WHEN RWY 13C WAS THE ACTIVE, AND NO CORRECTION BY TWR WHEN WE REPEATED WHAT WE THOUGHT WAS INSTRUCTION TO POS AND HOLD ON RWY 13L. I WOULD ASSUME THAT IT WAS STEPPED ON AS OTHER FLT RESPONDED SIMULTANEOUSLY. A HIGHER DEG OF SITUATIONAL AWARENESS BY US AND A NO ASSUMPTIONS ATTITUDE WOULD HAVE PREVENTED THIS SIT.

SYNOPSIS: THE FLC OF A B737 FINDS THEMSELVES NOSE TO NOSE ON TO A COMPANY FLT ON THE TKOF ROLL, OPPOSITE DIRECTION, ON RWY 31C AT MDW, IL.

Preparation and Remarks

The pilots making these reports were very familiar with their surroundings. There was no mention of confusion about runway/taxiway markings, signage, or lighting. It is assumed that they had and used a proper airport diagram.

Radio communications, other than being very busy, were normal to the operation until the time of the incident. Both pilots were listening to the controller's transmissions, and a read-back of the clearance was made. Note that the captain and the first officer were both aware of the possibility of an error due to similar aircraft call signs and challenged the controller on this issue. Both pilots were heads up—observing their surroundings.

Postincident Analysis

The incident was brought to a safe conclusion when the pilots and controller all noticed the problem, and the aircraft exited the runway. This incursion was classified as both pilot deviation and operational error. The incursion category would be category C because there was ample time and distance to avoid a potential collision. Note the comments appearing in the narrative: "A HIGHER DEG OF SITUATIONAL AWARENESS BY US AND A NO ASSUMPTIONS ATTITUDE WOULD HAVE PREVENTED THIS SIT."

Six specific problems are noted from the ASRS report:

Problem 1. Two aircraft had very similar call signs, causing confusion for the controller in issuing clearances and obtaining read-backs. The same problem existed for the aircraft crews. It was prudent of the flight crew to ask ATC for confirmation—although no response to the query was indicated.

Problem 2. The statements "TWR DID NOT RESPOND AND WE HAD NO INDICATION WE WERE BLOCKED" and "I WOULD ASSUME THAT IT WAS STEPPED ON AS OTHER FLT RESPONDED SIMULTANEOUSLY" indicate a possible explanation for some of the radio confusion. However, this was not known at the time by the flight crew.

Problem 3. "TWR DID NOT RESPOND" should have been followed with another immediate request for confirmation—but this does not appear to have been done.

Problem 4. Similar-sounding runway numbers—" "RUNWAY ONE THREE C" versus "RUNWAY THREE ONE C"—appeared to be in use at the same time. It is very easy to verbally confuse these two designations when issuing taxi instructions.

Problem 5. There was a clearance for an opposite-direction departure without notification.

Problem 6. There was an assumption by the flight crew for next release, but they did not get clearance for it.

Lessons Learned

This incident shows the absolute need for clear communications and the understanding of those communications between pilots and ATC.

Lesson 1. Both pilots must listen to all radio communications, not just those directed toward their specific call sign. Monitoring all the radio activity provides a good overview of who is where and doing what at the airport. In this situation, by actively monitoring all the radio traffic, the pilots were warned that all was not completely correct by the apparent confusion of call signs. However, ATC did not respond to their query. In a similar instance, it would be prudent to query ATC again and again until the call sign situation has been corrected.

Lesson 2. Keep the heads out of the cockpit when the aircraft is moving. In this particular incident, both pilots saw the landing lights of another aircraft taking off directly toward them from the opposite end of the runway. They took the prudent measure of exiting the runway.

Lesson 3. It would have been advisable, in this circumstance of call sign similarity and an unresponsive ATC, to have remained at the hold point until clarification from ATC was received.

Lesson 4. Turn on landing lights when taking off to provide as much of a visual footprint as possible for other aircrews. The lights seen at the far end of the runway were a strong warning that it was a good time to exit the runway.

Lesson 5. If you read back a clearance and get no response from the controller, challenge the controller until the read-back is acknowledged properly. Using this method may expend some air time, but it will prevent transmissions from becoming lost due to a blocked transmission (blocked is defined in Appendix A).

Lesson 6. The statement "ASSUMING THAT WE WOULD BE NEXT FOR RELEASE AND NOT GETTING POSITIVE CONFIRMATION OF THE CLRNC" is addressed in the second ASRS report for this incident when the pilot says that "A NO ASSUMPTIONS ATTITUDE" would have prevented the incident. Never assume anything—get it in fact before you take action.

Lesson 7. Controllers must be aware of communications problems arising from similar call signs of aircraft using the airport simultaneously and the similarity of runway numbers when said quickly. Clear enunciation is necessary—often helped by slowing down the speed of talking.

CASE 6

Better Signs

ASRS accession number: 454793

Month and year: November 1999

Local time of day: 1201 to 1800

Facility: MIA, Miami International Airport (FIG. 3-6)

Location: Miami, FL

Flight conditions: VMC

Aircraft 1: B737

Pilot of Aircraft 1: Captain, 24,300 hours; first officer

Reported by: Pilot

Incident description: Near runway incursion

Incident consequence: None

NARRATIVE AFTER PUSH BACK FROM GATE AT MIA, WE WERE CLRED TO TAXI TO RWY 9L VIA TXWY Q. WHEN REACHING THE INTERSECTION OF TXWY Q AND P, THE TAXI SIGN INDICATES TXWY P TO THE R AND TXWY Q TO THE L. TURNING TO THE L LINES YOU UP TO CONTINUE L ON TXWY T WHICH CROSSES RWY 12. I CONTINUED ON TXWY T AND ALMOST CROSSED RWY 12. I DID NOT ENTER THE RWY AND WAS ABLE TO HOLD SHORT. I TALKED TO THE TWR ABOUT THIS AND THEY SAID THAT THIS IS A PROB AREA AND FOUR ACR'S HAVE DONE THE SAME THING IN THE LAST 10 DAYS. BETTER SIGNS ARE NEEDED BEFORE SOMEONE GETS HURT.

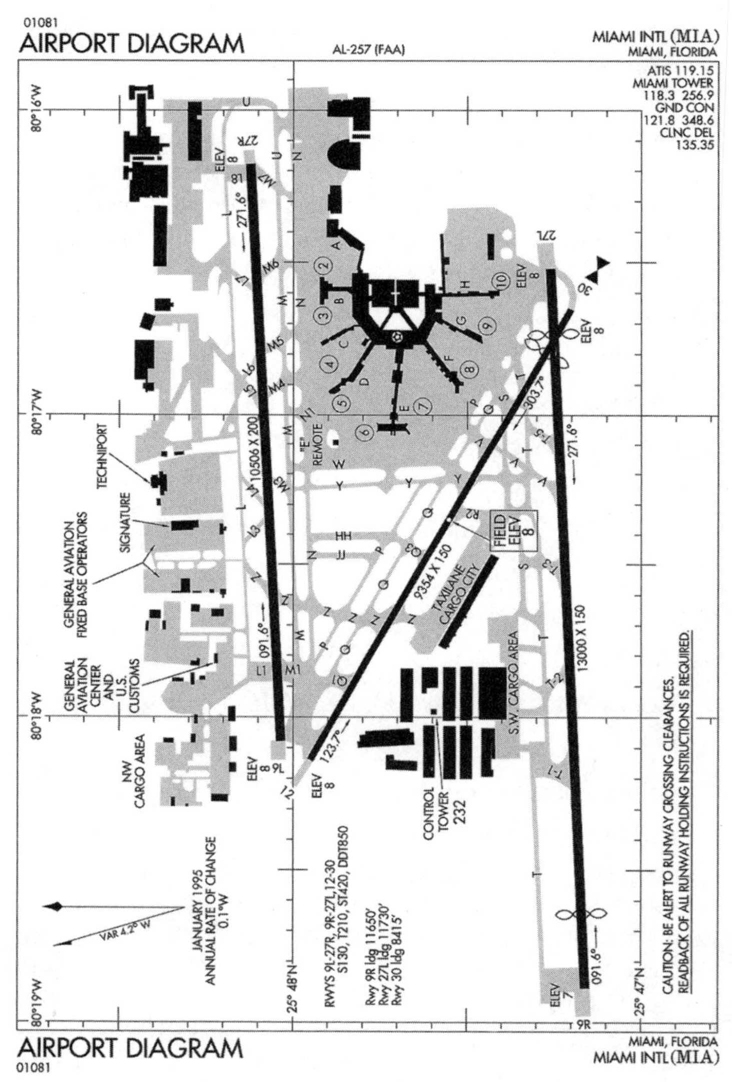

3-6 *Miami International, MIA (this airport diagram is not suitable for navigational purposes).*

SYNOPSIS B737-400 RPTR HAS DIFFICULTY WITH MIA SIGNAGE ON TAXI FROM GATE TO THE ACTIVE RWY.

Preparation and Remarks

The pilots making this ASRS are professional flight crew members. It is assumed that they are ready for the flight and adequately prepared and equipped with a proper airport diagram. Radio communications were normal, and there were no indications of problems in understanding the clearance given. There is no indication of anyone being rushed—controllers or flight crew. Normal preparation for receiving a taxi clearance calls for the airport diagram to be in hand—for visualizing purposes. There is no indication that this was done, although it would be proper procedure.

Postincident Analysis

No runway incursion resulted from this incident, but there could have been had the flight crew not been on its toes and stopped when it saw the hold-short surface markings for runway 12. The taxi path became confusing, and the aircraft wound up effectively lost on the airport, holding at a different runway from that intended.

Subsequent communication with the tower indicated that four similar incidents had occurred at the airport in the most recent 10 days. This incident, although not actually a runway incursion, is classified as an airport problem that is neither a pilot deviation nor an operational error. The incursion category would be category D because there was little or no chance of collision.

Two problem areas are noted specifically from this ASRS report:

Problem 1. There appears to be a problem with the airport signage indicating where specific taxiways are located. In this case, they are located at the intersection of taxiways Q, P, and T. The problem appears to be well known. Thus the airport would be responsible for taking corrective measures.

Problem 2. It does not appear that the flight crew reviewed the airport diagram while receiving taxi instructions or while taxiing. Had the flight crew visualized the ATC intended taxi path of taxiway Q to runway 9L, the aircraft probably would not have been turned so far left and wound up on taxiway T headed for runway 12.

Lessons Learned

We, as humans, depend on being given direction, either by direct message or by clues we look for. In this particular incident, it appears that there was a problem with taxiway signage and that ATC was aware of the problem, yet nothing had been done to correct the problem. However, signage aside, had adequate flight crew preparation been made, it is possible the incident would never have happened.

Lesson 1. Airport signage must be placed correctly so that the signs' messages leave nothing to be misunderstood. Just as "clear and concise" are essential when talking on the radio, the signs must be clear and concise in their meaning.

Lesson 2. The use of airport diagrams while receiving taxi instructions and while taxiing can greatly reduce the tendency to become lost on the airport. In this particular incident, tracing the taxi instruction out on the airport diagram probably would have prevented the incident. As an aid to using runway diagrams, check your aircraft's heading in reference to the diagram and where you think you should be, where you want to be, and the path between the two.

CASE 7

Unnoticed Incursion

ASRS accession number: 460626

Month and year: January 2000

Local time of day: 1201 to 1800

Facility: PBI, Palm Beach International Airport (FIG. 3-7)

Location: West Palm Beach, FL

Flight conditions: VMC

Aircraft 1: Mooney M-20J

Pilot of aircraft 1: Single pilot, 550 hours

Reported by: Pilot

Incident description: Runway incursion

Incident consequence: FAA reviewed incident with pilot.

NARRATIVE AT XA50 WAS GIVEN CLEARANCE TAXI RWY 9L, WAS GIVEN PROGRESSIVE TAXI INSTRUCTIONS, THEN TOLD TO TAXI END OF TAXIWAY R RUNUP AREA. FOLLOWING INSTRUCTIONS THERE APPEARED TO BE A RUNUP AREA AFTER HAVING CROSSED WHAT APPEARED TO BE BLACK PAINTED OVER RWY HOLD BAR. RWY 9L WAS CLEARLY OVER 100 YARDS AWAY 'NORTH'. AFTER MY RUNUP, I CONTACTED TOWER AND WAS CLEARED RWY 9R DEPARTURE. AFTER TURNING THE PLANE 120 DEGREES R I DISCOVERED MY RUNUP WAS ACTUALLY DONE ON RUNWAY 9R. I DID NOT OBSERVE ANY SIGNAGE FROM TAXIWAY R OF RWY 9R AND ONLY OBSERVED GROUND MARKINGS FOR RWY 9R. WHILE I WAS NOT NOTIFIED BY GROUND OR TOWER OF MY INCURSION THERE ARE SIGNIFICANT SAFETY CONCERNS OF GROUND INSTRUCTIONS, FIELD OBSERVATION BY TOWER, SIGNAGE, AND MARKING OF RWY 9R FROM TAXIWAY R. NO OTHER AIRCRAFT WERE USING 9R.

SYNOPSIS MOONEY PLT INCURS RWY APT PBI.

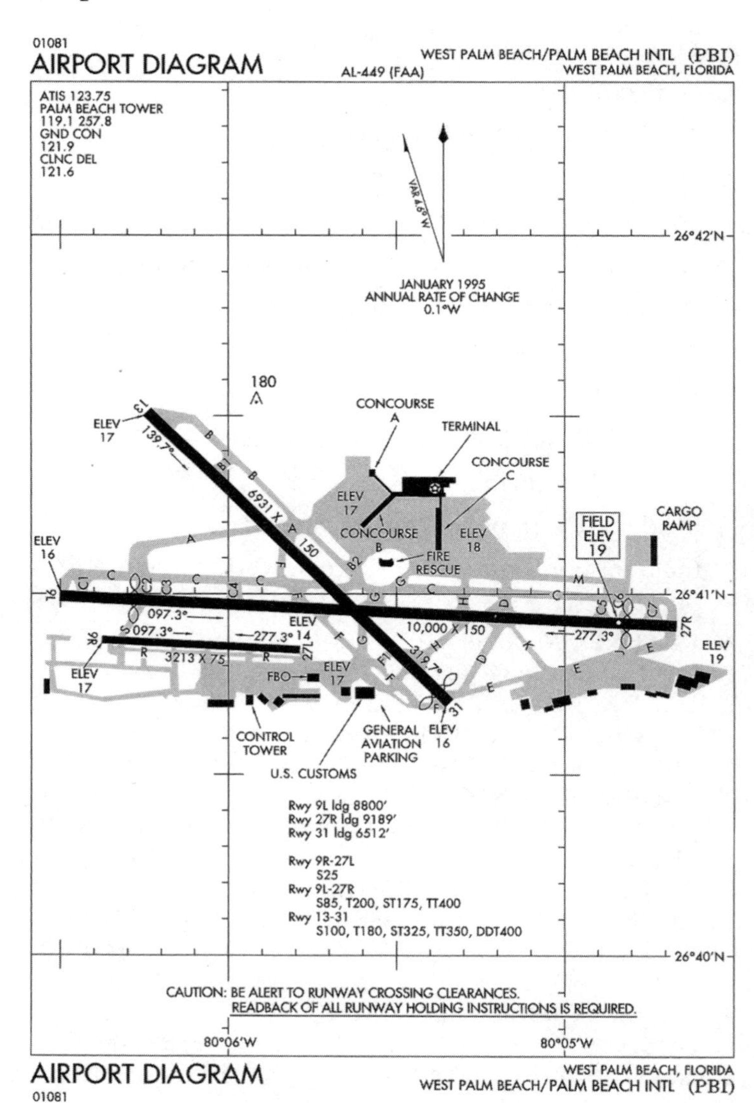

3-7 *Palm Beach International, PBI (this airport diagram is not suitable for navigational purposes).*

Preparation and Remarks

The pilot making the report indicated that there were no specific preincident problems and there appeared to be nothing confusing about the taxi clearance received. The pilot did indicate receiving progressive taxi instructions. It is doubtful the pilot was familiar with the airport, and there is no indication of airport diagram use. Radio communications were normal.

Postincident Analysis

This incident went undetected by the tower controller and was reported later by the pilot. No separation was lost. However, an incursion did occur.

Following taxiway R was not a problem for the pilot, but locating the run-up area was. Follow along the airport diagram for Palm Beach International Airport, and you can see that the aircraft left from the FBO area and traveled in a westerly direction along taxiway R. At the west end of the taxiway is the run-up area—indicated as much larger in size than the width of the taxiway. The run-up area exits to runway 9R.

In order for the pilot to place the aircraft on runway 9R, a 90-degree right turn had to be made from taxiway R or from the run-up area. A pilot using an airport diagram would have seen that the entry to the runway required a 90-degree turn from the taxiway—the taxiway being parallel to the runway in question for its entirety.

The report states, "I DID NOT OBSERVE ANY SIGNAGE FROM TAXIWAY R OF RWY 9R AND ONLY OBSERVED GROUND MARKINGS FOR RWY 9R." This brings two questions:

1. Were there no signs, or were the signs merely not observed?
2. If the ground markings for runway 9R were seen (as stated), why was there an incursion?

The statement "APPEARED TO BE A RUNUP AREA AFTER HAVING CROSSED WHAT APPEARED TO BE BLACK PAINTED OVER RWY HOLD BAR" can be analyzed as meaning that the taxiway markings appearing black actually were the runway holding position markings (at the entrance to runway 9R), possibly reflecting oddly due to the angle of the sun at that particular time of day.

This runway incursion would be classified as pilot deviation—the aircraft was where it should not have been. The category of the incursion would be category D because there was little or no chance of a collision, but an incursion did occur.

In addition to the problem of an incursion caused by pilot deviation is the fact that the controller completely missed the incident. As the report states, where was the "FIELD OBSERVATION BY TOWER"?

The specific problems noted from this ASRS reported incident include what was written in the report and what does not appear in the report:

Problem 1. The pilot indicated that there were no marking or signs observed indicating where the taxiway ended and the runway began.

Problem 2. The pilot was not observant enough to notice the turn made prior to entering the runway—a turn that is typical of many intersections of taxiways and runways at this and most other airports.

Problem 3. There were visual clues available before or as the incursion was made. The clues include the general appearance of a runway as compared with a taxiway, as well as the fact that surface markings on a runway differ from those on a taxiway.

Problem 4. The tower controller missed observing the incursion—which could have created a very serious incident.

Lessons Learned

While reviewing the ASRS report of this incident, it is unclear if the pilot was aware of the number of errors made while taxiing and doing a run-up, although blame is placed as follows: "WHILE I WAS NOT NOTIFIED BY GROUND OR TOWER OF MY INCURSION THERE ARE SIGNIFICANT SAFETY CONCERNS OF GROUND INSTRUCTIONS, FIELD OBSERVATION BY TOWER, SIGNAGE, AND MARKING OF RWY 9R FROM TAXIWAY R." There are several lessons to be learned from this incident.

Lesson 1. Use an airport diagram when receiving a taxi clearance, and use it while you taxi. This will help in reducing confusion about where you are and where you are going. It also will provide clues as to where you should be looking for surface markings and/or signage.

Lesson 2. The tower controller apparently was not vigilant of aircraft activity on the airport. Otherwise, the incursion would have been seen. Perhaps, had the airplane been followed visually by the controller, the incursion might have been prevented by a communication from the tower.

Lesson 3. The pilot must be heads up and ever vigilant about the surroundings—not counting on others to do the job. The pilot is in command and is responsible for the aircraft.

Lesson 4. Anytime there is a question about signage, markings, or lighting on runways and/or taxiways, contact ATC at the facility involved. Note the name(s) of person (s) contacted and include that contact information in an ASRS report.

CASE 8

Controller Talks Too Fast

ASRS accession number: 455600

Month and year: November 1999

Local time of day: 0601 to 1200

Facility: DFW, Dallas–Fort Worth International Airport (FIG. 3-8)

Location: Dallas–Fort Worth, TX

Flight conditions: VMC

Aircraft 1: B737

Aircraft 2: MU-2

Pilot of aircraft 1: Captain; first officer, 9600 hours

Pilot of aircraft 2: Single pilot

Reported by: Pilot of aircraft 1

Incident description: Runway incursion

Incident consequence: FAA reviewed incident with pilot.

NARRATIVE. DFW TWR CLRED US FOR TKOF RWY 35L. I TOOK CTL OF THE ACFT AFTER ACFT WAS ALIGNED ON CTRLINE. I NOTICED AN MU2 TAXIING L TO R AND SAID 'WATCH THAT MU2.' I BELIEVE IT WAS ON TXWY A. THE MU2 CROSSED RWY 35L W TO E. I VERIFIED WITH TWR THAT WE WERE CLRED FOR TKOF. AGAIN, WE WERE. THE MU2 WAS WELL CLR OF RWY 35L AND WE BEGAN OUR TKOF ROLL (AFTER CAPT HAD NOTIFIED TWR OF WHAT WE SAW). CALLBACK CONVERSATION WITH RPTR REVEALED THE FOLLOWING INFO: CREW WAS FLYING A B737-500. THE MU2 WAS ALSO ON THE TWR FREQ. THE RPTR NOTED THE MACHINE GUN RAPIDITY WITH WHICH THE TWR CTLR WAS ISSUING CLRNCS. HE DESCRIBED IT AS AUCTIONEER FAST. HE THINKS THAT CTLRS SHOULD BE COUNSELED UNTIL SPEECH RATE IS SLOW ENOUGH TO BE UNDERSTANDABLE BY THE PLTS INVOLVED IN THE OP. HE ATTRIBUTES THE FAILURE OF THE MU2 TO HOLD SHORT TO THE CTLR'S SPEECH RATE.

SYNOPSIS B737 CREW HAD MU2 CROSS RWY IN FRONT OF THEM AFTER THEY HAD BEEN CLRED FOR TKOF.

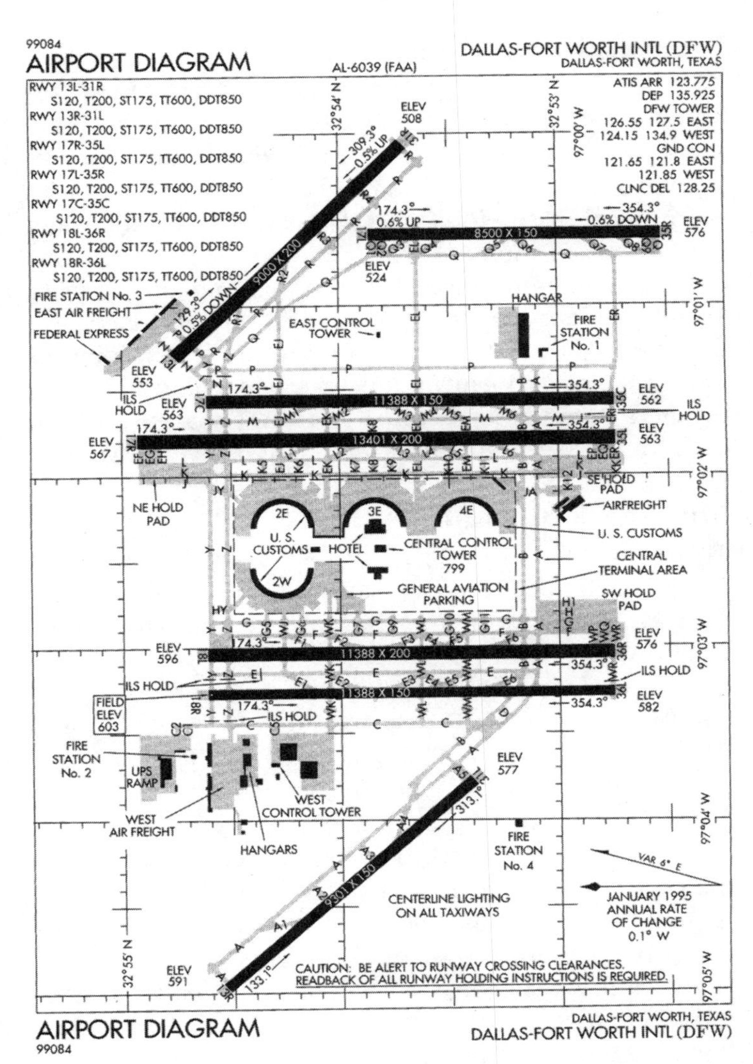

3-8 *Dallas-Fort Worth International, DFW (this airport diagram is not suitable for navigational purposes).*

Preparation and Remarks

The pilot making the ASRS report was aware of the surroundings. There had been no apparent confusion on the flight crew's part about the clearance, and all was proceeding normally until the aircraft had been lined up with the runway just before takeoff. At that point, the flight crew observed another aircraft crossing the active runway.

Radio communications for the departing aircraft were normal, except as noted in the report. All indications are that the flight crew was prepared for and was doing its job as expected—including visual vigilance (which resulted in the clearance query to ATC based on their observation of an aircraft crossing the active runway).

Postincident Analysis

This incident was brought to a safe conclusion by an observant flight crew that saw the MU-2 crossing the runway ahead of them—the runway on which they (B737) were cleared for takeoff.

The reporting pilot indicated in the report "THE MACHINE GUN RAPIDITY WITH WHICH THE TWR CTLR WAS ISSUING CLRNCS. HE DESCRIBED IT AS AUCTIONEER FAST" as being the prime cause for the MU-2's incursion across the active runway.

The incursion caused the B737 to delay takeoff until the controller had been advised of the situation. For this reason, the runway incursion severity category would be category C because evasive action required. The classification of the incursion, however, is not as easy to determine. The ASRS report indicates both ATC and the MU-2's flight crew were deficient in human performance.

Several specific problems are noted from the ASRS report:

Problem 1. The report does not address whether the MU-2 was cleared to cross the runway in error or whether the MU-2 missed a clearance. From the report, it is understood that the MU-2 may have received what its flight crew thought was a clearance to cross runway 35L.

Problem 2. There is no indication that a read-back of the taxi clearance was made (by either aircraft).

Problem 3. The report indicates that the controller was talking too fast to be easily understood. Although this may appear to be a likely place of blame, remember that the pilot in command has the responsibility for the aircraft [FAR 91.3 (a)]: "The pilot in command of an aircraft is directly responsible for, and is the final authority as to, the operation of that aircraft."

Problem 4. When the MU-2 started to cross runway 35L, did that pilot check for aircraft on the runway? Assuming that the B737 had its landing lights on, as required for takeoff, they should have been visible to the MU-2.

Problem 5. Where was ATC when this happened? Was the controller visually observing the aircraft under ATC control?

Lessons Learned

This incident shows how very important it is for clear and concise communications on the parts of both pilots and controllers.

Lesson 1. The alert flight crew in the B737 saw the incursion and checked with the tower to be sure they were in fact cleared for takeoff. Once again, the flight crew's heads-up attitude saved the day.

Lesson 2. It is quite possible that during the reported "AUCTIONEER FAST" high-speed controller instructions, either the MU-2 failed to receive a hold-short instruction or did not understand it. A simple read-back of the taxi instruction to the controller might have prevented the incident.

Lesson 3. Read-back of hold-short instructions is necessary to emphasize the clearance in the pilot's mind as well as to be a cross-check for errors [AIM Section 4-3-18(a)8].

Lesson 4. Clear and concise communications are necessary for ATC to work properly. Speak clearly, using standard terms (see AIM Pilot/Controller Glossary), and question anything that is not perfectly clear. Speed talking only compounds the problem of understandability—leading to mistakes.

Lesson 5. When having trouble understanding a controller, use either "speak slower," "say again," or "words twice" to require the controller to speak in an understandable manner. These phrases appear in the Pilot/ATC Glossary in Appendix A. Use of any of these phrases will remind the controller to slow down.

CASE 9

Too Many Directions

ASRS accession number: 427904

Month and year: February 1999

Local time of day: 1201 to 1800

Facility: CLE, Cleveland-Hopkins International Airport (FIG. 3-9)

Location: Cleveland, OH

Flight conditions: VMC

Aircraft 1: Beech King Air 300

Pilot of aircraft 1: PIC, 3200 hours

Reported by: Pilot

Incident description: Runway incursion

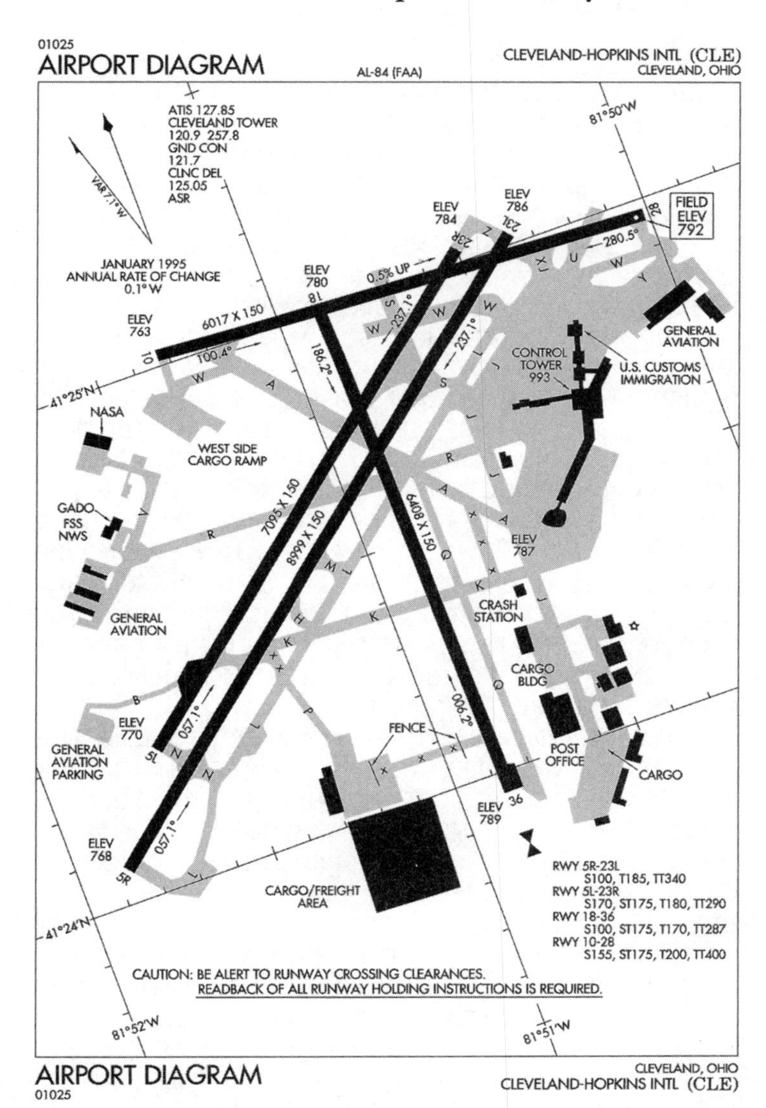

3-9 *Cleveland-Hopkins International, CLE (this airport diagram is not suitable for navigational purposes).*

Incident consequence: None

NARRATIVE THE REASON I AM FILING THIS RPT IS BECAUSE WE CROSSED AN ACTIVE RWY WITHOUT CLRNC. THE MISUNDERSTANDING HAPPENED WHILE TAXIING AT CLE. THE CTLR SAID TO TAXI ON TXWYS L, W, J, RWY 28, HOLD SHORT OF RWY 23L, EXPECT RWY 23R FOR DEP. THIS IS WHAT WE HEARD: TAXI NE ON TXWYS L, W, RWY 28, RWY 23R, EXPECT RWY 23R FOR DEP. THIS IS WHAT WE ACTUALLY DID: NE ON TXWY L TURNED L ON TXWY W, CROSSED RWY 23L AND HELD SHORT OF RWY 23R. IF YOU LOOK AT THE ARPT DIAGRAM, TXWY W IS VERY CONFUSING AND IT LOOKS LIKE FROM TXWY L YOU CAN ONLY TURN L. I FEEL THAT THE CONTRIBUTING FACTORS TO THIS PROB ARE: 1) THAT THE CTLR GAVE US TOO MANY INSTRUCTIONS ALL AT ONCE 2) THE TXWYS ARE VERY HARD TO UNDERSTAND. FINALLY, I THINK THAT I SHOULD HAVE STUDIED THE DIAGRAM MORE AND SHOULD HAVE QUESTIONED ATC BEFORE PROCEEDING ANY FURTHER. I THINK TO HELP PREVENT THIS FROM HAPPENING IN THE FUTURE IS TO TAKE A LOOK AT RENAMING SOME OF THE TXWYS. ALSO, TO SHORTEN UP CLRNCS THAT ARE GIVEN OUT BY ATC.

CALLBACK CONVERSATION WITH RPTR REVEALED THE FOLLOWING INFO: CREW WAS FLYING A BEECH KING AIR 300. CREW WERE STRANGERS AT THE CLE ARPT. THE CAPT STATED THAT THE GND CTLR WAS BUSY AND GAVE A LONG CLRNC AT A VERY FAST SPEECH RATE. WHEN THE CAPT GOT TO THE NE END OF TXWY L, THE TXWY SIGNS BECAME CONFUSING TO HIM. HE STATED THAT IT APPEARED THE ONLY WAY TO TURN WAS L. HE SAW THE HOLD SHORT LINES FOR RWY 23L BUT THOUGHT HE WAS CLR TO CROSS THE RWY. AT NO POINT DID HE ASK FOR 'PROGRESSIVE' TAXI INSTRUCTIONS. ON DISCUSSION WITH CLE FSDO, FAA STATED THAT THE HOLD BAR ON RWY 28 WAS ADDED BECAUSE OF THE MANY RWY INCURSIONS HAPPENING THERE. SINCE THIS ADDITION, AND INCREASED SIGNAGE, AND AN EDUCATIONAL EFFORT WITH THE ACRS OPERATING THERE, THE RWY INCUR-

SIONS HAVE GONE AWAY. THE ONLY HOLD FOR RWY 23L IS THE STANDARD HOLD PAINT MARKINGS. THE RWY IS MARKED WITH STANDARD RWY SIGNS.

SYNOPSIS BE30 CPR CREW BECAME CONFUSED WITH THE GND CTLR TAXI INSTRUCTIONS AT CLE. HE MAY HAVE HAD A RWY INCURSION.

Preparation and Remarks

The pilots making this ASRS report were new to Cleveland-Hopkins International Airport and not familiar with its operations. There was reported confusion about the taxi clearance (quite lengthy) and mention of a need to study the airport diagram more closely. From the statement, it is assumed that the pilot was using an airport diagram. Radio communications appear to have been normal, with no malfunctions noted.

The facility has had a problem with runway incursions in the recent past and now has an in-place training program for regular users. Further, a hold bar for runway 28 had been added.

Postincident Analysis

This incident was brought to a safe conclusion because there was no loss of separation during the incursion. Safe conclusion was not brought about by the actions either of an alert flight crew or of the tower controller.

This incursion incident was caused by pilot deviation, which resulted from total confusion with the taxi instructions received. The pilot felt, in defense, that too many instructions were given at one time and that the airport taxiway system is confusing. However, in the report, the pilot offered, "I THINK THAT I SHOULD HAVE STUDIED THE DIAGRAM MORE AND SHOULD HAVE QUESTIONED ATC BEFORE PROCEEDING ANY FURTHER."

There is no indication that the controller was visually observing airport activity; otherwise, there may have been some corrective action by the controller.

There were no other aircraft involved with this incursion: therefore, the runway incursion severity category would be category D because there was little or no chance of a collision. However, the incident still meets the definition of a runway incursion.

A number of problems are noted from this incident:

Problem 1: The taxi clearance was rather long and apparently given rather quickly by the controller. This may have initiated the incident, but there is no indication that the pilot read back the taxi instructions, asked for clarification, asked for a repeat, or requested progressive instructions.

Problem 2. The pilots were confused by the clearance and the airport diagram.

Problem 3. At no time was the controller challenged about the clearance—as shown by the statement, "AT NO POINT DID HE ASK FOR 'PROGRESSIVE' TAXI INSTRUCTIONS."

Problem 4. There may be a problem with confusing signage—something pilots unfamiliar with CLE would not be accustomed to.

Problem 5: ATC has recognized that there is a problem at CLE and has an ongoing education program at the local pilot level. A similar educational program should be directed toward the controllers so that there will be less confusion about clearances as they become aware of the problems facing pilots using CLE.

Problem 6. The controller did not visually observe the incident.

Lessons Learned

The primary lessons learned from this incident deal with the failure of the flight crew to ask questions when they became confused. The action of the aircraft is the responsibility of the pilot in command [FAR Section 91.3 (a)].

Lesson 1. When you are unsure of a taxi instruction or clearance (or anything else involving ATC and aircraft movement) given by a controller, ask for a repeat. Then give the controller a read-back.

Lesson 2. Check the airport diagram as you receive your taxi clearance. Visualize it. If it does not visualize easily, the chances are things will go no better when you start moving.

Lesson 3. If there is still confusion about the taxi clearance, request "progressive taxi instructions." This will cause the controller to walk you through each step of the taxi operation—allowing very little room for confusion or error.

Lesson 4. The controller may have been talking too fast, causing the pilots to have difficulty in understanding the clearance. This does not present a problem if you just ask the controller to "speak slower" or "say again" the clearance. The controller will slow down the second, third, or fourth time the clearance must be repeated.

Lesson 5. Slow down—for all involved. Make sure that you understand the clearance you receive; ask for repeats as necessary, and do not move until you are sure you are ready. The rest of the world will wait for you—perhaps begrudgingly so—but wait they will. And in the long run, you will get to your destination anyway. Get hurried, and you may not get there at all.

Lesson 6. The local flight standards district office (FSDO) at CLE has the right idea. It recognized a problem and has an ongoing educational program to correct it. This program should be continued into the cab of the tower to create an atmosphere of deliberate cooperation on the part of clearance-issuing controllers.

CASE 10

Language Barrier

ASRS accession number: 431312

Month and year: March 1999

Local time of day: 1201 to 1800

Facility: BOG, Bogotá International Airport (no diagram)

Location: Bogota, Columbia

Flight conditions: Marginal

Aircraft 1: B727

Aircraft 2: Unknown

Pilot of aircraft 1: Captain; first officer, 6500 hours; second officer

Pilot of aircraft 2: Unknown

Reported by: Pilot(s) of aircraft 1

Incident description: Runway incursion

Incident consequence: Aircraft 2 made a go-around.

NARRATIVE WHILE WAITING #1 FOR TKO, THE BOG TWR ISSUED THE FOLLOWING CLRNC: 'LINE UP AND WAIT RWY 13L.' ALL 3 PLTS IN THE COCKPIT UNDERSTOOD THE SAME THING. I, AS PNF, RESPONDED 'LINE UP AND WAIT RWY 13L.' WE TOOK THE RWY AND WAITED IN POS. THE CAPT AND MYSELF LOOKED ON FINAL, AND WE DID NOT SEE ANYBODY. LATER THE TWR TOLD US THAT 'THE CLRNC WAS AFTER LNDG TFC, LINE UP AND WAIT.' WE RESPONDED THAT WE DID NOT UNDERSTAND AFTER

LNDG TFC BECAUSE WE NEVER HEARD IT. THE CTLR SENT THE AIRLINER IN A GAR. THE LANGUAGE BARRIER AND DIFFERENT PHRASEOLOGY USED OUTSIDE THE UNITED STATES MIGHT HAVE BEEN A CONTRIBUTING FACTOR FOR THE INCURSION.

SYNOPSIS. A FLC ACCEPTS A CLRNC ONTO A RWY BUT DOES NOT HEAR THE CONDITION OF 'AFTER THE LNDG ACFT, LINE UP AND WAIT.' FLC FORCES LNDG ACFT TO GO AROUND AFTER THEY MOVE INTO POS FOR TKO.

Preparation and Remarks

Takeoff operations, including the reception of a takeoff clearance, were normal until the tower indicated a runway incursion had happened, forcing a landing aircraft to make a go-around. Note that all three pilots were monitoring the radio communications with the tower, as shown in the report by, "ALL 3 PLTS IN THE COCKPIT UNDERSTOOD THE SAME THING."

Postincident Analysis

The reason for this runway incursion, according to the ASRS report filed by the pilot, probably was the language and phraseology barrier. Slight differences in ATC operations also may have contributed.

The report states, "WE TOOK THE RWY AND WAITED IN POS. THE CAPT AND MYSELF LOOKED ON FINAL, AND WE DID NOT SEE ANYBODY." A landing aircraft was not seen and was not monitored on the radio as landing.

The classification for this incursion is pilot deviation because the aircraft entered a runway without clearance. The runway incursion severity category is category B because separation was decreased and there was a significant potential for collision. In this case, the landing aircraft was instructed to make a go-around by an observant tower controller.

Although the flight crew blames the BOG controller, a language barrier, and minor differences in ATC procedures, two problems were noted when reading AIM Section 4-3-18 and comparing it with this report:

Problem 1. The instruction "LINE UP AND WAIT RWY 13L" did not specify exactly where to wait—at runway 13L or on runway 13L. However, in accordance with AIM Section 4-3-18 (7), this would have been a valid taxi instruction requiring the aircraft to wait at runway 13L. Specifically, the AIM says in part, "This does not authorize the aircraft to 'enter' or 'cross' the assigned departure runway at any point."

Problem 2. The tower apparently implied "AFTER LNDG TFC" but did not specifically so state. There is no need to imply—the clearance was to runway 13L, *not onto* runway 13L.

Lessons Learned

This runway incursion may have been prevented had the tower's instructions been somewhat more clear and/or the flight crew had understood that local operations may imply certain meanings within the taxi clearances. However, when comparing the AIM with what occurred during this incursion, the following lessons are noted:

Lesson 1. The original clearance appears to have been quite clear: Taxi to runway 13L—emphasis being on the word *to*. When in doubt about any clearance, ask the controller to "say again" the clearance. Be sure you understand it completely. An alleged language barrier is no defense for failure to comply with ATC instructions.

Lesson 2. The International Civil Aviation Organization (ICAO) has created terminology standards for ATC (see the AIM Pilot/Controller Glossary). The same terminology is used worldwide—as an aid to standardization of airport operations.

Lesson 3. When there is a doubt as to what the clearance means, query the controller.

CASE 11

Where Are You?

ASRS accession number: 453492

Month and year: October 1999

Local time of day: 1201 to 1800

Facility: CAK, Akron-Canton Regional Airport (FIG. 3-10)

Location: Akron, OH

Flight conditions: VMC

Aircraft 1: Learjet 35

Aircraft 2: Learjet 24

Pilot of aircraft 1: Captain

Pilot of aircraft 2: Captain

Reported by: Instructor controller

Incident description: Near runway incursion

Incident consequence: None

NARRATIVE ACFT #1 WAS GIVEN A TAXI CLRNC TO RWY 23 AFTER HE RPTED HIS POS WAS AT FBO-1. FBO-1 IS AN FBO AT CAK ARPT. IT IS LOCATED ON THE E SIDE OF THE AIRFIELD. THE APCH END OF RWY 23 IS ALSO LOCATED ON THE NE END OF THE FIELD. THERE ARE NO OTHER RWYS TO CROSS BTWN FBO-1 AND RWY 23. TXWY E PARALLELS THE RAMP AREA THAT LEADS INTO FBO-1 AND IS THE TXWY THAT LEADS TO THE APCH END OF RWY 23. ANOTHER ACFT THAT WAS GIVEN TAXI INSTRUCTIONS FROM FBO-1 TO RWY 23, WAS INSTRUCTED TO FOLLOW ACFT #1 (LR35) TO RWY 23. AT THIS TIME ACFT #1 INDICATED HE WAS ON THE W SIDE

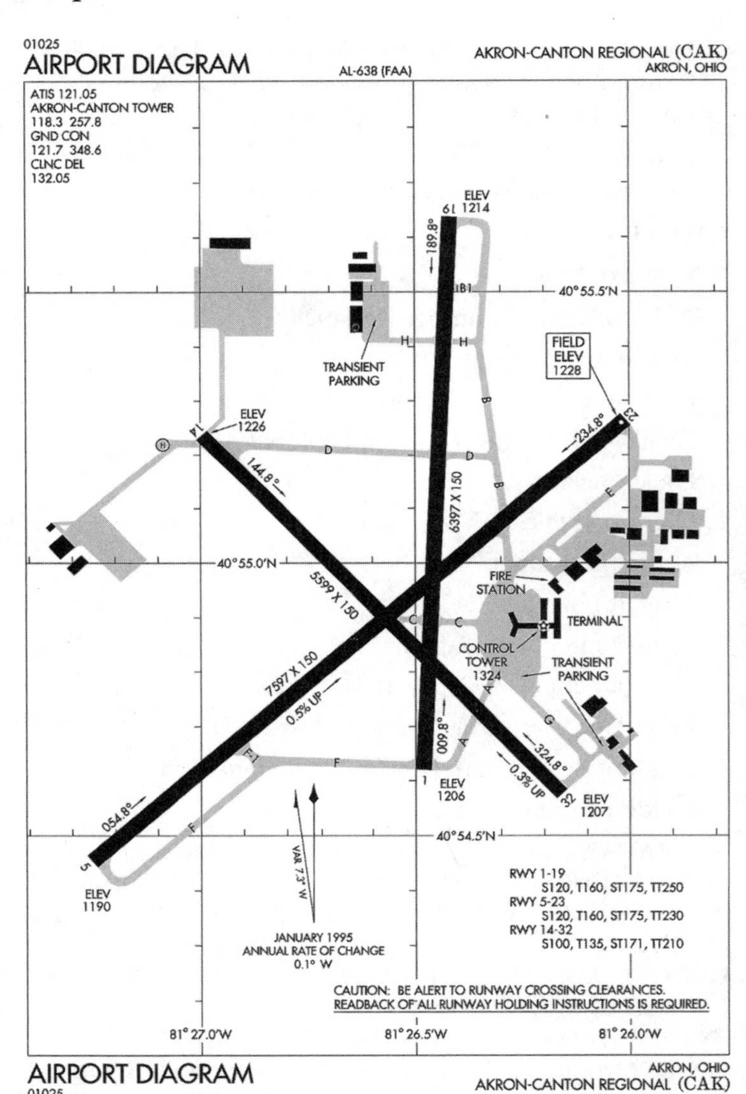

3-10 *Akron-Canton Regional Airport, CAK (this airport diagram is not suitable for navigational purposes).*

OF THE ARPT. ACFT #1 WAS INSTRUCTED IMMEDIATELY TO 'HOLD SHORT OF RWY 19 AT TXWY H.' ACFT #1 WAS OBSERVED JUST ENTERING ONTO TXWY H. HOWEVER, THERE WAS ANOTHER LEARJET (ACFT #2) DEPARTING RWY 19. BECAUSE OF ACFT #1'S ERRONEOUS POS RPT AND BECAUSE OF THE CAK ARPT'S MGMNT FAILURE TO TAKE ACTION ON THE FOLLOWING, THE POTENTIAL EXISTS FOR A CATASTROPHE. THERE WAS NO LOSS OF SEPARATION THIS TIME, BUT I BELIEVE THE POTENTIAL EXISTS EVERYDAY! FBO-1 IS AN FBO ON THE E SIDE OF CAK ARPT. THIS IS USUALLY REFERRED TO AS FBO E. THIS IS BECAUSE AT ONE TIME, FBO-1 WAS ALSO LOCATED ON THE WNW SIDE OF THE FIELD. THIS ONE NO LONGER EXISTS. THE FBO ON THE W SIDE NOW IS 'FBO-2.' IT IS MY UNDERSTANDING ACFT USING 'FBO-2' ARE INSTRUCTED TO SAY 'FBO-2' WHEN ASKED THEIR POS, AND THERE ARE ALSO SIGNS INSTRUCTING THEM ALSO. HOWEVER, IT IS MY UNDERSTANDING THERE ARE SIGNS WITH 'FBO-1' ON THEM, STILL IN VIEW ON THE W SIDE.

THESE 'OLD' SIGNS MISLEAD PLTS AS TO THEIR POS AND AS IN THE CASE BE GIVEN WRONG TAXI INSTRUCTIONS.

SYNOPSIS CAK DEVELOPMENTAL CTLR RECOGNIZES TAXIING ACFT HAS RPTED TAXIING FROM AN INCORRECTLY IDENTED ARPT FBO AND CORRECTS CLRNC TO HOLD SHORT OF RWY FOR DEPARTING TFC.

Preparation and Remarks

The flight crew of aircraft 1 was not aware of any problems until a taxi clearance was issued for another aircraft to follow their aircraft. Perhaps the flight crew noticed that it was not being followed by aircraft 2, or perhaps it felt that the taxi clearance issued to their aircraft was too simple, since from the airport diagram the clearance should have involved several taxiways and the crossing of two runways. Airport radio communications were normal.

Postincident Analysis

An actual runway incursion was prevented when that aircraft's position on the airport was brought to the controller's attention by a radio transmission, as stated on the ASRS report: "AT THIS TIME ACFT #1 INDICATED HE WAS ON THE W SIDE OF THE ARPT." At that time "ACFT #1 WAS INSTRUCTED IMMEDIATELY TO 'HOLD SHORT OF RWY 19 AT TXWY H.'"

It is obvious that the controller did not know where aircraft 1 was because the original taxi instruction did not include crossing any runways—as would be required to get to runway 23 from the FBO on the west side of the airport.

There is no incursion classification or category for this incident because there was no actual runway incursion and no loss of separation resulted from this incident. The example only serves to show how minor confusion can lead to major incidents.

Several problems involving the controller, airport signage, and local customs are involved in this incident:

Problem 1. There is no indication of a taxiway clearance being issued that included any taxiway designations. Had there been, such as taxiway E, it would have been immediately clear to the flight crew of aircraft 1 that there was a problem in the controller's recognition of their location on the airport. The controller's assumption of location and the issuance of "TAXI CLRNC TO RWY 23" indicate a failure by the controller to communicate with the aircraft. The controller should have queried the aircraft as to its actual position on the airport.

Problem 2. The controller did not initially visually observe aircraft 1, or there would have been no confusion.

Problem 3. The flight crew in aircraft 2 should have been monitoring the radio transmissions (no mention of this in the report) and should have reported that there was no aircraft to follow.

Problem 4. The reporting person has an issue with the use of the airport signs indicating FBO locations and how flight crews are to announce themselves to ATC.

Problem 5. This type problem generally only affects flight crews not familiar with CAK's layout and operation.

Lessons Learned

The lessons learned from this incident affect flight crews, controllers, and CAK's airport management and serve to demonstrate the amount of cooperation required to make airport operations safe.

Lesson 1. Use airport diagrams. When you receive a taxi clearance, check the clearance with the airport diagram to see if it works. If something is questionable about the clearance, ask the controller. In this case, aircraft 1 should have asked about crossing runway 19.

Lesson 2. The controller should observe where aircraft are located on the ground when issuing taxi clearances. If you do not see the aircraft, you are depending on the flight crew to know where they are—and they may not know (as was the case in this example).

Lesson 3. Airport management should remove all confusing signs and replace them with clear, concise, and easily read signs—signs that are designed to impart a specific message in a very easy and very accurate manner.

CASE 12

Follow the Leader

ASRS accession number: 456361

Month and year: December 1999

Local time of day: 1201 to 1800

Facility: CLE, Cleveland-Hopkins International Airport (FIG. 3-11)

Location: Cleveland, OH

Flight conditions: VMC

Aircraft 1: B757

Aircraft 2: Unknown

Pilot of aircraft 1: First officer, 10,000 hours; captain

Pilot of aircraft 2: Unknown

Reported by: First officer of aircraft 1

Incident description: Runway incursion

Incident consequence: Landing aircraft did a go-around.

NARRATIVE AT CLE, WAS ASKED BY GND CTL TO FOLLOW ACFT AHEAD AND MONITOR TWR. WE SWITCHED TO TWR, THEY THEN ADVISED ACFT AHEAD TO 'TAXI ACROSS RWY 23L, INTO POS AND HOLD RWY 23R.' I BELIEVE THE CAPT WAS STILL THINKING HE WAS TO FOLLOW THAT ACFT. SO, AS HE PROCEEDED ACROSS THE HOLD SHORT LINE FOR RWY 23L, I MENTIONED THAT 'I THINK YOU'RE SUPPOSED TO STOP BEHIND THE HOLD LIGHT BAR.' AT THAT SAME MOMENT THE TWR SENT AN ACFT AROUND AND ADVISED US WE WERE ON AN ACTIVE RWY. I ESTIMATE 10 FT ACROSS THE LINE WHEN HE STOPPED.

SYNOPSIS A B757 FLT PIC TAXIES ACROSS THE HOLD LINE FOR RWY 23L WHILE 'FOLLOWING TFC' AT CLE, OH.

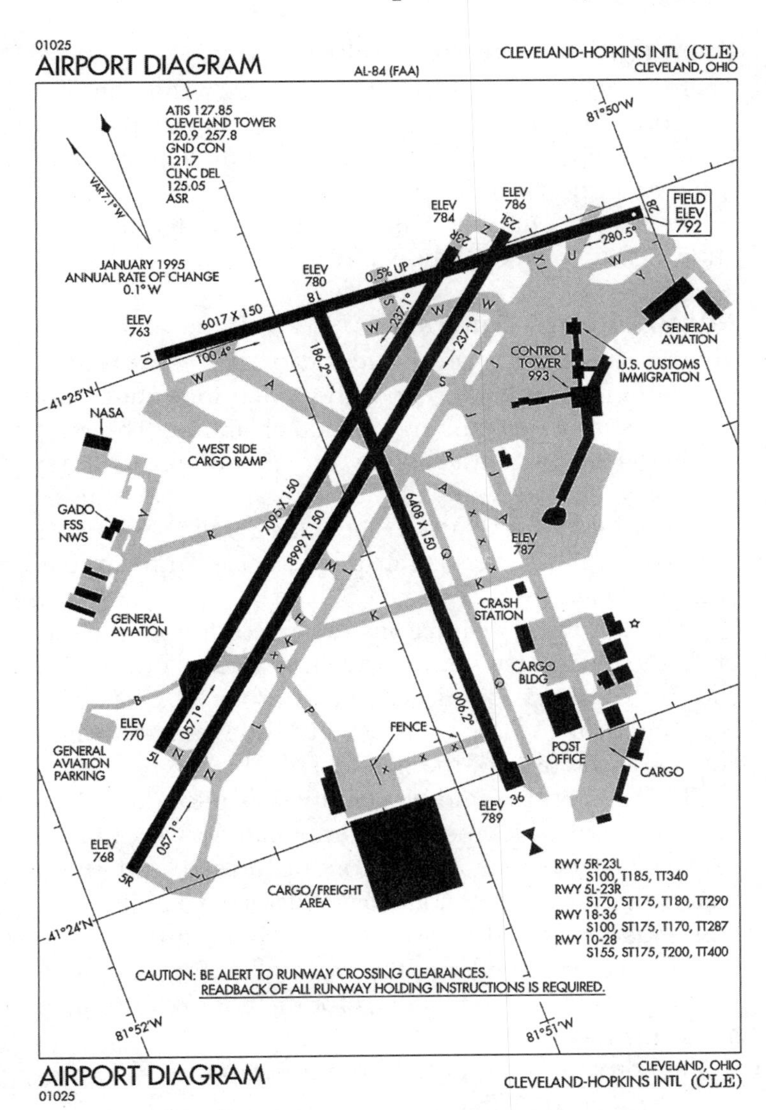

3-11 *Cleveland-Hopkins International Airport, CLE (this airport diagram is not suitable for navigational purposes).*

Preparation and Remarks

There appeared to be no problems with radio communications, and the flight crew was having no problem knowing where they were. It is assumed they were using an airport diagram.

The tower controller was alert and corrected the situation before any separation was lost.

Postincident Analysis

There is no mention in the report of any conversation between the controller and the pilot indicating any problems in understanding the taxi clearance. The pilot of aircraft 1 was told to follow the aircraft ahead. Later the aircraft ahead received a clearance to "TAXI ACROSS RWY 23L, INTO POS AND HOLD RWY 23R." Aircraft 1 started to follow the aircraft ahead across runway 23L without clearance.

Although the initial taxi instruction of follow the aircraft ahead is somewhat vague, it is understood that a runway cannot be crossed without specific clearance. Not stated in the report is whether the hold-bar light was on at the time of the incident.

This incursion would be classified as pilot deviation, and the runway incursion severity category would be category B because separation decreased and there was a significant potential for collision. A landing aircraft was instructed to go around. There also is some blame to be placed on the controller for issuing a vague instruction (easy means for the controller to give a taxi instruction).

Two problems are noted from this report:

Problem 1. The controller must be more precise than just instructing an aircraft to follow the aircraft ahead—unless there is a direct taxi path with no holds or stops. Stating the taxi instruction as, "Aircraft xxx taxi to runway 23L behind aircraft xxx and hold" would have been more precise. Aircraft xxx would then read back the hold instruction.

Problem 2. The pilot of aircraft 1 should have queried the tower when the lead aircraft received a clearance to cross runway 23L to ascertain if he also was cleared to cross runway 23L.

Lessons Learned

You can take nothing for granted in ATC situations. Always make sure you completely understand what is said—and what is meant. If there is room for doubt, then there is also a need to query the controller.

Lesson 1. Never cross a hold line unless you have expressly been cleared to do so. You are in violation of the FARs if you do.

Lesson 2. When there are any doubts about a clearance, stop before passing any hold signs or surface markings and query the controller (never stop on a runway).

CASE 13

Displaced and Unseen

ASRS accession number: 457049

Month and year: December 1999

Local time of day: 0601 to 1200

Facility: IPT, Williamsport Regional Airport (FIG. 3-12)

Location: Williamsport, PA

Flight conditions: VMC

Aircraft 1: Learjet 35

Pilot of aircraft 1: Captain, 18,000 hours; first officer,

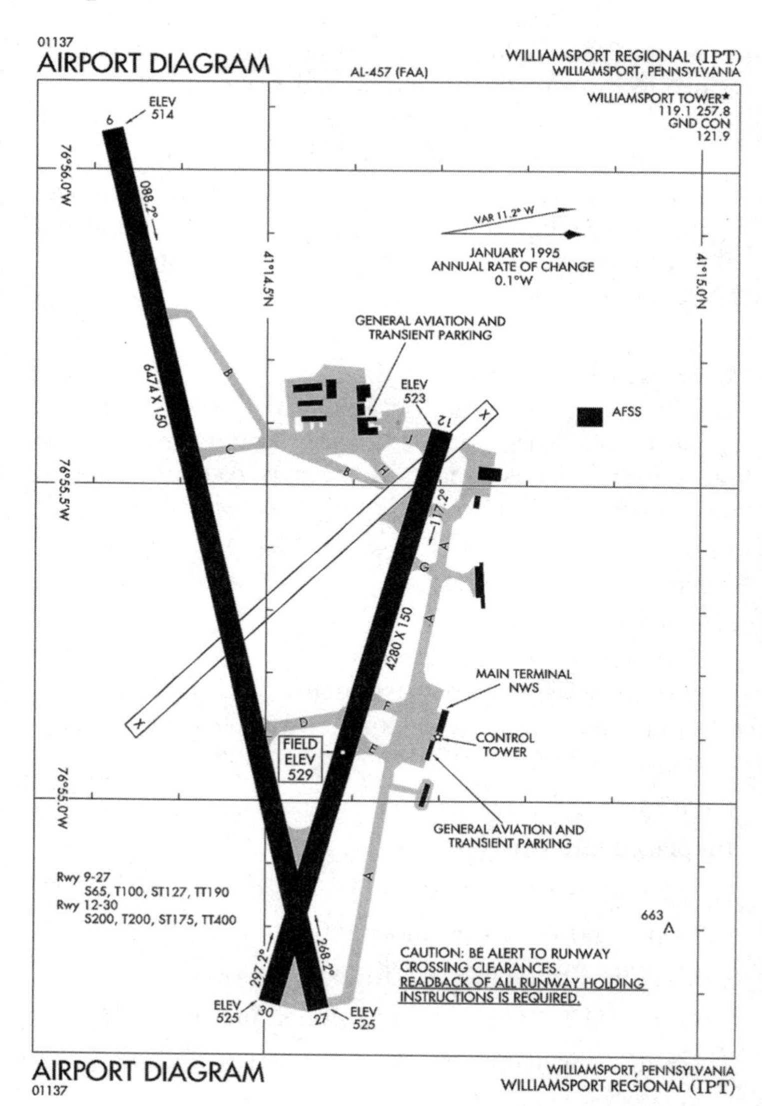

3-12 *Williamsport Regional, IPT (this airport diagram is not suitable for navigational purposes).*

9500 hours

Reported by: Pilot of aircraft 1

Incident description: Runway incursion

Incident consequence: FAA reviewed the incident with the flight crew.

NARRATIVE OUR ACFT (X) OVERRAN THE HOLD SHORT LINE FOR RWY 27 AT WILLIAMSPORT, PA (IPT). WE WERE CLRED BY GND CTL FROM THE RAMP TO RWY 27. THIS INVOLVED CROSSING INACTIVE RWYS. WE VERIFIED BY A SECOND CALL THAT WE WERE CLRED TO CROSS AND PROCEED TO RWY 27 WITHOUT A HOLD SHORT. THE WEATHER WAS 10 MI VISIBILITY AND CLR OF CLOUDS. THERE WERE NO OTHER ACFT TAXIING OR OPERATING IN THE PATTERN. WE HAD COMPLETED THE TAXI CHECKLIST IN CHALLENGE AND RESPONSE STYLE JUST AS WE APCHED RWY 27. AT THIS TIME, GND CTL POLITELY INFORMED US THAT THE RWY HAD A DISPLACED HOLD LINE AND INDEED THERE WAS NO LINE AT THE RWY'S BORDER AS ONE MIGHT HAVE EXPECTED. THERE WAS NO TFC, AND THE TWR CLRED US ONTO THE RWY WITHOUT FURTHER MENTION. AS WE WERE NEW TO THIS FIELD, WE HAD MADE A GOOD STUDY OF A COMMERCIAL CHART OF ARPT LAYOUT, AND NO SPECIAL NOTES EXIST TO POINT OUT THIS UNUSUALLY EXAGGERATED HOLD SHORT POINT. WHILE WE ARE SURE THE HOLD LINES ARE PAINTED ON THE TXWY, THIS POINT ON TXWY A SHOULD BE MADE HIGHLY VISIBLE TO PLTS DUE TO ITS UNUSUAL LOCATION. CREWS THAT ARE COMPLETING CHECKS IN COMPLICATED ACFT WILL CERTAINLY OPERATE SAFER WITH EFFECTIVE NOTIFICATION OF NONSTANDARD PROC. FROM THE TWR'S RESPONSE, WE FELT THAT THIS HAPPENS OFTEN AT IPT. NOTATION ON THE ARPT DIAGRAM AS WELL AS EFFECTIVE SIGNAGE WOULD HAVE HELPED THIS SIT.

SYNOPSIS LJ-35 FLC CROSSED RWY 27 HOLD SHORT LINE, STATING DISPLACE THRESHOLD CONDITION DOES NOT REFLECT RWY HOLD MARKINGS WHERE ONE WOULD EXPECT SUCH MARKINGS.

Preparation and Remarks

The statement from the report, "WE HAD MADE A GOOD STUDY OF A COMMERCIAL CHART OF ARPT LAYOUT, AND NO SPECIAL NOTES EXIST TO POINT OUT THIS UNUSUALLY EXAGGERATED HOLD SHORT POINT," says it all. The flight crew was prepared as much as possible. No problems with radio communications with ATC were noted, and there were no other aircraft operating at the airport at the time of the incident.

Postincident Analysis

The incident was caused by the flight crew not seeing the displaced hold line for runway 27 and thus crossing it. The location of the line is somewhat premature from the point normally expected for a hold line.

The incursion would be classified as a pilot deviation, and the runway incursion severity category would be category D because there was little or no chance of a collision. However, it meets the definition of a runway incursion. The aircraft did pass the hold line.

The problems noted from this incident involve signage and paying attention outside the aircraft:

Problem 1. The statement, "WE HAD COMPLETED THE TAXI CHECKLIST IN CHALLENGE AND RESPONSE STYLE JUST AS WE APCHED RWY 27. AT THIS TIME, GND CTL POLITELY INFORMED US THAT THE RWY HAD A DISPLACED HOLD LINE," from the report indicates that there was too much inside activity and not enough heads up. Note that the controller did observe the incursion and pointed it out to the flight crew.

Problem 2. Although not noted on the airport diagram, the hold line is not in a location normally associated with hold lines. Therefore, some type of signage should be installed to draw pilots' attention to the line.

Lessons Learned

Taxiing always requires outside observation—not just cursory glances when it is convenient for the flight crew.

Lesson 1. Keep your heads up, and look outside. This single effort will eliminate most runway incursions.

Lesson 2. The airport should place obvious and attention-gaining signs at the location of the displaced hold line.

CASE 14

Runway Taxi

ASRS accession number: 465940

Month and year: March 2000

Local time of day: 0601 to 1200

Facility: STL, Lambert–St. Louis International Airport (FIG. 3-13)

Location: St. Louis, MO

Flight conditions: VMC

Aircraft 1: B737

Pilot of aircraft 1: Captain, 10,000 hours; first officer

Reported by: Pilot of aircraft 1

Incident description: Runway incursion

Incident consequence: None

NARRATIVE TAXI CLRNC WAS 'TAXI VIA RWY 6. HOLD SHORT OF RWY 12L.' RWY 6 DOES NOT ACTUALLY

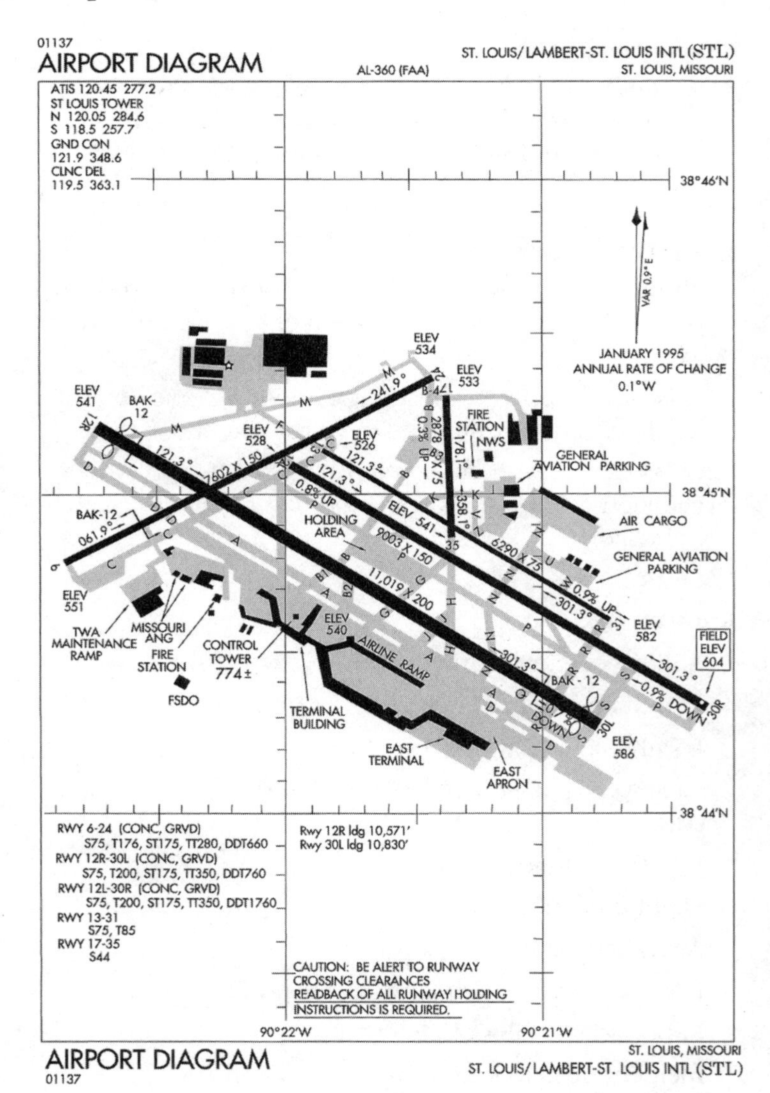

3-13 *St. Louis International, STL (this airport diagram is not suitable for navigational purposes).*

INTERSECT RWY 12L, BUT DOES INTERSECT THE APCH LIGHTS OF RWY 12L. I WAS NOT PAYING CLOSE ENOUGH ATTENTION AND INADVERTENTLY WENT BEYOND THE HOLD SHORT LINE ABOUT A PLANE LENGTH. I DISCOVERED THIS AS I CROSSED THE LINE AND THE FO WAS TELLING ME OF THAT FACT. THE APCH PATH WAS CLR, AS RWY 12L WAS BEING USED FOR DEPS. SO AS FAR AS I KNOW THERE WAS NO CONFLICT. TWR SUBSEQUENTLY MANEUVERED US TO THE OTHER SIDE SO WE WERE LEGAL AGAIN. IT IS A SLIGHTLY DIFFERENT THAN NORMAL SIT TO TAXI VIA A RWY TO HOLD SHORT OF THE APCH PATH, BUT I HAVE LEARNED TO BE EXTRA VIGILANT NOW WHEN THINGS ARE ABNORMAL.

SYNOPSIS B737 CREW HAD A RWY INCURSION ON RWY 12L AT STL.

Preparation and Remarks

The incident involved taxiing on a runway—which is not normal. Although the flight crew was properly prepared for operations at STL, this abnormality upset the routine. There were no problems noted with radio communications.

Postincident Analysis

The incident was caused by the flight crew not stopping at the hold-short line for runway 12L (actually an area prior to the runway threshold). The cause of the problem was clearly stated in the report as, "I WAS NOT PAYING CLOSE ENOUGH ATTENTION AND INADVERTENTLY WENT BEYOND THE HOLD SHORT LINE ABOUT A PLANE LENGTH. I DISCOVERED THIS AS I CROSSED THE LINE AND THE FO WAS TELLING ME OF THAT FACT."

This incursion was due to pilot deviation. The runway incursion severity category would only be category D—because there was little or no chance of a collision on this area of pavement.

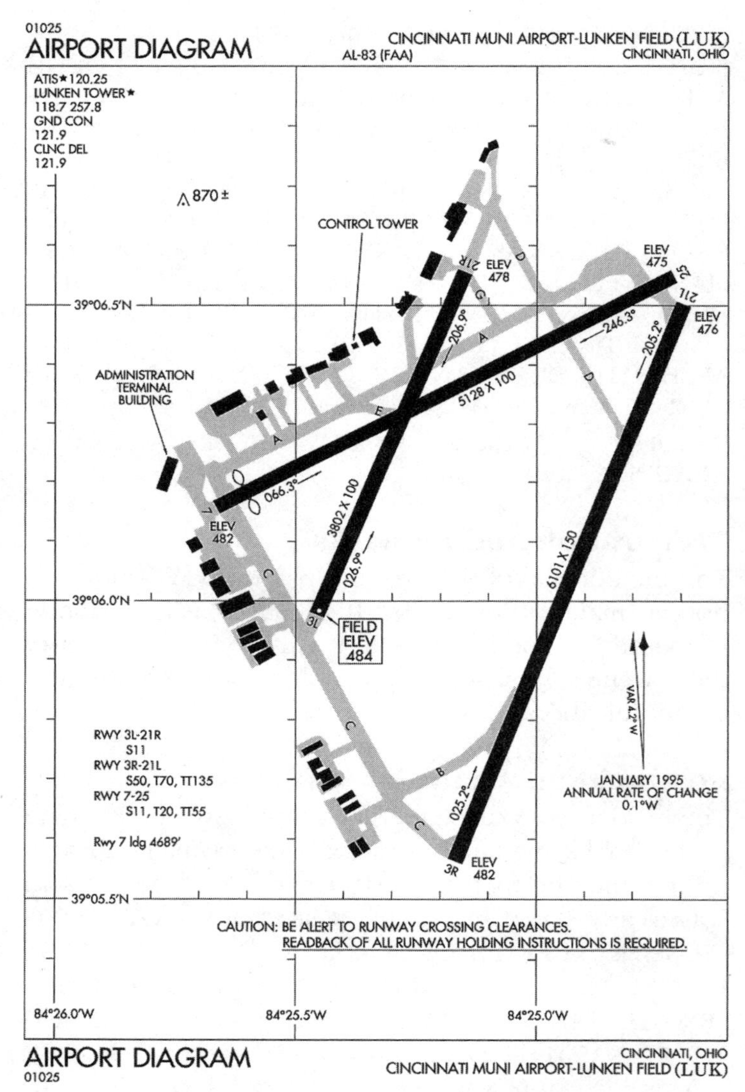

3-14 *Cincinnati Municipal Airport, LUK (this airport diagram is not suitable for navigational purposes).*

The problem noted with this incursion is specifically a lack of observation on the part of the pilot during an abnormal taxi (taxiing on a runway).

Problem 1. Lack of pilot observation.

Lessons Learned

Taxiing always requires outside observation, and when taxiing on a runway, which is not normal, even more attention to the outside must be made.

Lesson 1. Keep your heads up and look outside.

CASE 15

Chip on the Shoulder

ASRS accession number: 454330

Month and year: November 1999

Local time of day: 0601 to 1200

Facility: LEK, Cincinnati Municipal Airport–Lunken Field (FIG. 3-14)

Location: Cincinnati, OH

Flight conditions: VMC

Aircraft 1: Piper PA-31 Navajo

Pilot of aircraft 1: Captain, 9000 hours; first officer

Reported by: Pilot of aircraft 1

Incident description: Runway incursion

Incident consequence: Issued advisory

NARRATIVE CLRED TO TAXI FROM FBO TO RWY 21L VIA TXWY C AND TXWY A HOLD SHORT OF RWY 21R. AS WE APCHED THE EXTENDED CTRLINE OF THE DEP END OF RWY 21R THE CTLR ADVISED US TO STOP IMMEDIATELY AND REMINDED US THAT WE WERE INSTRUCTED TO HOLD SHORT OF RWY 21R AND THAT THERE WAS A SIGN ON THE TXWY THAT POINTED THIS OUT. AS ONE CAN SEE BY THE ENCLOSED ARPT DIAGRAM, RWY 21R

DOES NOT INTERSECT WITH TXWY C. THEREFORE, ONE CANNOT HOLD SHORT OF RWY 21R ON TXWY C, BUT ONE CAN HOLD SHORT OF THE EXTENDED CTRLINE OF RWY 21R ON TXWY C, IF INSTRUCTED TO DO SO.

SYNOPSIS PA31 CREW DOES NOT HOLD SHORT AS INSTRUCTED IN TAXI CLRNC.

Preparation and Remarks

Talk about an inefficient taxiway system. This airport has that. To taxi from the southeastern service areas to runway 21L is nearly a case of "can't get there from here." However, the pilot was prepared for the lengthy taxi and was using an airport diagram. No problems were noted with radio communications.

Postincident Analysis

In reading the report, there appears to be a problem in semantics. The controller meant for the Piper to hold at the extended centerline of runway 21R, but the exact statement was not made. The aircraft was told to hold at runway 21R—which should have meant the same to the pilot. Runway and taxiway hold instructions are not something to pick at. They are intended to maintain aircraft separation—not be examples of perfect word usage.

This was a pilot deviation, and the runway incursion severity category would be category D because there was no apparent loss of separation or danger of collision.

The problem associated with this reported incident appears to be a pilot looking for an argument:

Problem 1. The pilot knew, or should have known, what the controller meant by saying, "HOLD SHORT OF RWY 21R." If the pilot did not understand this statement, perhaps some remedial training would be in order. Otherwise, an attitude check would do fine.

Lessons Learned

This incident serves to show that there are those among the ranks of pilots and controllers just looking for excuses to not follow instructions and to cause trouble when questioned about their actions.

Lesson 1. Do what the controller says—and if there is a slightly incorrect usage of a term or a phrase, let it go. It probably was not intentional, and the message did get across, so there is no problem. Further, there is no point in upsetting your day or the day of the controller by showing how knowledgeable you are.

Lessons Learned in this Chapter

This chapter has closely examined a number of different departure-type runway incursions—as reported on the ASRS reports included with the case studies. Pilot deviations and operational errors were shown. Recognize that flight crews can do little about operational errors, except for vigilance and bringing these errors to the attention of ATC. These cases demonstrate the need for full involvement by flight crews and controllers in the reduction or prevention of runway incursions.

In each case study, several lessons were learned. Of these lessons, the following appear to be the most common:

Visual vigilance. For flight crews, keep your heads out of cockpits, and look around—watch for other aircraft activity on the airport. For controllers, be sure to know where the aircraft with which you are working are—not every pilot reports a correct position.

Radio vigilance. By listening to all ATC's radio traffic, not just traffic to you alone, you will know what is going on around you at all times. You will have the "big picture."

Airport diagrams. Airport diagrams can prevent you from becoming lost on the airport. They provide a means to visualize a taxi clearance as it is given to you by ATC. Additionally, referencing a diagram while taxiing can prevent entering a runway by mistake (but remember to keep your head out of the cockpit).

Read back taxi clearances. All taxi clearances containing hold instructions must be read back. Read-back is an excellent means of avoiding both pilot deviations and operational errors.

As a pilot, anytime you have a problem, either with a taxi instruction, a clearance as received, or while taxiing, contact ATC for help. If you do not understand a clearance, say either "say again" or "words twice" to get a repeat.

From AIM Section 4-3-18 (b): "ATC clearances or instructions pertaining to taxiing are predicated on known traffic and known physical airport conditions." Therefore, it is important that pilots clearly understand the clearance or instruction. Although an ATC clearance is issued for taxiing purposes, when operating in accordance with CFRs, it is the responsibility of the pilot to avoid collision with other aircraft. Since "the pilot-in-command of an aircraft is directly responsible for, and is the final authority as to, the operation of that aircraft," the pilot should obtain clarification of any clearance or instruction that is not understood.

4

Arrival Runway Incursions

Runway incursions during arrival operations occur as an aircraft is landing or about to land. The aircraft in question can be on the ground or in the air. This means that speeds during the incident are normally significantly faster than those found during departure incursions.

In this chapter, the Aviation Safety Reporting System (ASRS) reports examined for the case studies all resulted from incidents at towered (controlled) airports. Some of the incidents involved pilot error, which the Federal Aviation Administration (FAA) refers to as pilot deviation resulting from the violation of any Federal Aviation Regulations (FARs). Other incidents came from operational errors, those attributed to air traffic control (ATC). Still others stem from vehicle/pedestrian deviations.

In each type of runway incursion—those caused by pilot deviations, those caused by operational errors, and those caused by vehicle/pedestrian deviations—there were actions that could have been taken to prevent the incident from occurring.

CASE 1

Blocked Radio Transmissions

ASRS accession number: 456445

Month and year: December 1999

Local time of day: 0601 to 1200

Facility: LAX, Los Angeles International Airport (FIG. 4-1)

Location: Los Angeles, CA

Flight conditions: Marginal

Aircraft 1: Wide body, low wing, 3 turbojet engines

Aircraft 2: Unknown

Pilot of aircraft 1: First officer; captain, 8300 hours

Pilot of aircraft 2: PIC unknown

Reported by: Pilot of aircraft 1

Incident description: Runway incursion

Incident consequence: FAA reviewed incident with pilot.

NARRATIVE ON LNDG AT LAX RWY 25L, WE WERE GIVEN INSTRUCTIONS TO TURN L OFF OF RWY, TAXI DOWN PARALLEL AND HOLD SHORT OF RWY 25L ON TXWY F. WHEN WE ARRIVED AT TXWY F, WE WERE TOLD TO CROSS RWY 25L AND THAT THE CTLR WOULD CALL US BACK, AND THAT TFC WAS ON A 1½-MI FINAL. THE FO READ THE CLRNC BACK AND NOTHING FURTHER WAS SAID. UPON REACHING RWY 25R WE WERE TOLD TO CROSS RWY 25R AND JOIN TXWY B. WHEN I ARRIVED AT THE HOTEL, I HAD A MESSAGE TO CALL THE CTL TWR. THE SUPVR INFORMED ME THAT WE DID NOT HAVE CLRNC TO CROSS RWY 25L AND THAT A RPT WOULD BE FILED. THE SUPVR SAID THE TAPE OF THE CTLR STATED 'EXPECT TO CROSS RWY 25L AWAIT FURTHER INSTRUCTIONS.' THE TFC WAS VERY BUSY THAT DAY AND THERE WERE A LOT OF RADIO XMISSIONS BEING BLOCKED BY OTHER TFC. I BELIEVE THAT THE 'EXPECT TO' PART OF

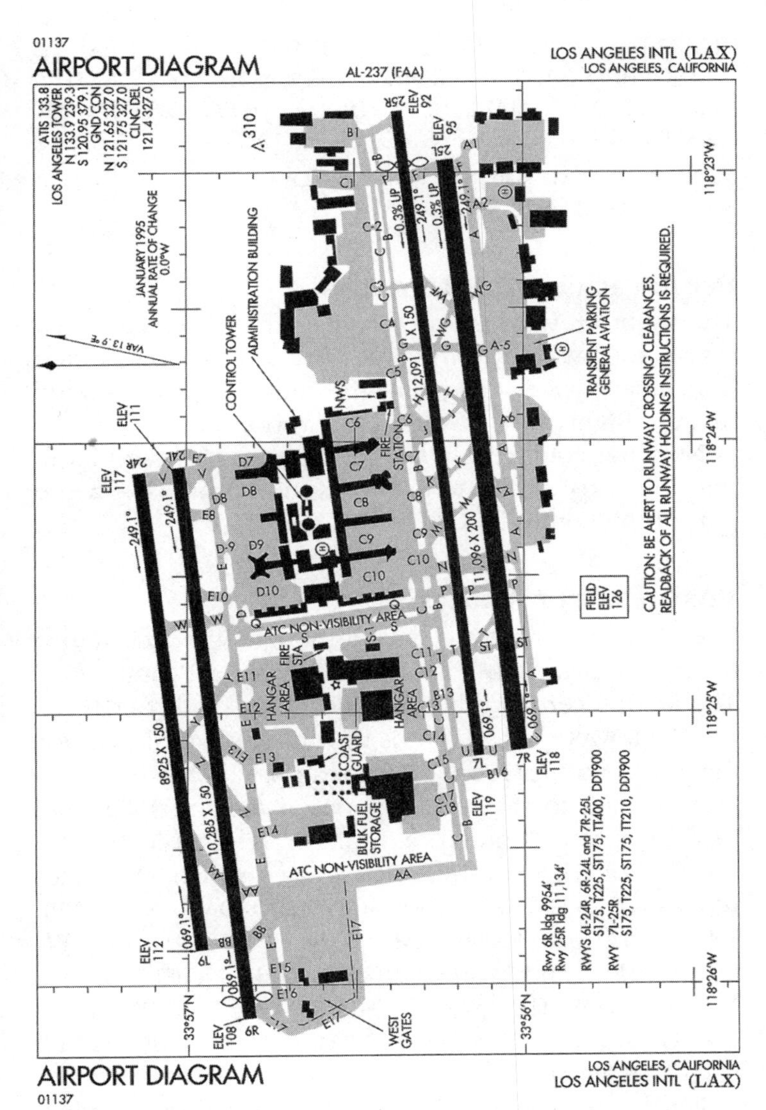

4-1 *Los Angeles International, LAX (this airport diagram is not suitable for navigational purposes).*

THE XMISSION WAS BLOCKED AND WE DID NOT HEAR IT. I ALSO BELIEVE THE PROPER TERMINOLOGY FOR THIS CLRNC SHOULD HAVE BEEN 'HOLD SHORT OF RWY 25L AWAIT FURTHER INSTRUCTIONS, TFC ON 1½-MI FINAL.'

SYNOPSIS ACR CROSSES RWY AT LAX AFTER MISUNDERSTANDING CLRNC.

Preparation and Remarks

The flight crew received landing and preliminary taxi instructions with no problems. There were no reported difficulties while navigating the airport. It is assumed that the flight crew was using an airport diagram.

Note the comment in the report about blocked radio transmissions and a terminology use problem on the part of the controller.

Postincident Analysis

The flight crew received what it thought was a clearance to cross runway 25L—with an aircraft on a 1½-mile final. The first officer read the clearance back, and "NOTHING FURTHER WAS SAID." There is no indication on this report that the controller received the read-back.

According to the ATC supervisor, the controller had said (as recorded on tape), "EXPECT TO CROSS RWY 25L AWAIT FURTHER INSTRUCTIONS," to which the pilot replies, "THE TFC WAS VERY BUSY THAT DAY AND THERE WERE A LOT OF RADIO XMISSIONS BEING BLOCKED BY OTHER TFC" and "I ALSO BELIEVE THE PROPER TERMINOLOGY FOR THIS CLRNC SHOULD HAVE BEEN 'HOLD SHORT OF RWY 25L AWAIT FURTHER INSTRUCTIONS, TFC ON 1½-MI FINAL.'"

For the purpose of examining this case, there is only a single tape recording of the incident—that of the controller. This recording does not show what was received—or what was blocked—at the aircraft. Therefore, problems with garbled and blocked transmissions become a matter of conjecture. Note, however, that cockpit voice recorders (CVRs) can be used to settle questions about radio reception at the aircraft. Blocked transmissions will not be heard by the CVR any differently than by the flight crew.

The incident is classified as pilot deviation, and the report does indicate human performance problems with the crew and ATC. The runway incursion severity category for this incident is category C because separation decreased, but there was ample time and distance to avoid a potential collision. No go-around was ordered for the landing aircraft.

Several problems are noted with this incident:

Problem 1. LAX is a very busy airport, and as indicated in the report, radio transmissions are prone to being blocked. If there is any doubt about an ATC instruction, ask the controller to "say again" the clearance—either all of it or the part you indicate.

Problem 2. A clearance to cross an active runway with another aircraft on a 1½-mile final for the same runway should raise a warning flag. There is not enough time for the taxiing aircraft to cross the runway and not cause a loss of separation. The flight crew should have challenged the controller on this clearance and asked the controller to "say again."

Problem 3. There is no indication the controller ever received the read-back. It would be appropriate to ask the controller if the read-back was received. You could make the read-back and preface it with, "Please confirm my read-back of …"

Problem 4. The term *expect* is not normally used in the context used by this controller. Further, the term *await* is not in the Pilot/Controller Glossary. However, a simple hold-short instruction will stop the aircraft until further instructions are issued.

Problem 5. There is nothing in the report about the crew looking before crossing the runway. If an aircraft were landing, its landing lights should have been on—thereby making it relatively easy to see.

Lessons Learned

This seemingly simple incident is actually a culmination of several small operational errors and pilot deviations. It offers several lessons:

Lesson 1. Be visually vigilant. Look before you cross a runway or taxiway. If you observe something that may cause a conflict or loss of separation, ask the controller to clarify instructions to cross.

Lesson 2. Clearance across an active runway should cause extra vigilance—both visually and while monitoring the radio. Monitoring the radio allows the pilots to know where other traffic is and what it is doing.

Lesson 3. At a busy airport, there is little that presently can be done (technology-wise) about the radio transmission blockage problem. There is just simply too great a volume of radio traffic at any point in time on the same frequency. To combat the chance of a lost, incomplete, or questionable clearance, give a read-back—and be sure that the controller acknowledges it.

Lesson 4. Standard terminology should be used by all involved with ATC. This means not just pilots, but members of the ATC team as well. This will reduce confusion, which may have been a part of the problem in this incident.

CASE 2

Controller Blames the System

ASRS Accession number: 451812

Month and year: October 1999

Local time of day: 0601 to 1200

Facility: SJC, San Jose International Airport (FIG. 4-2)

Location: San Jose, CA

Flight conditions: VMC

Aircraft 1: Cessna 150

Pilot of aircraft 1: Single pilot

Reported by: Controller

Incident description: Runway incursion

Incident consequence: None

NARRATIVE ACFT INSTRUCTED TO MAKE R TFC RWY 29. PLT ACKNOWLEDGED AND SUBSEQUENTLY WAS CLRED RWY 29. PLT READ BACK INSTRUCTIONS AND CLRNC CORRECTLY. PLT LANDED RWY 30R. CONTRIBUTING FACTOR: UNFAMILIAR WITH ARPT ENVIRONMENT. THIS WAS EVIDENT TO GND CTL WHEN PLT DID NOT FOLLOW INSTRUCTIONS. POSSIBLE RWY MARKINGS. FROM CTLR'S VIEW, PLT SOUNDED AS IF HE WAS FAMILIAR. DID NOT QUESTION CLRNC, WHICH LED THIS CTLR TO PAY LITTLE OR NO ATTN ONCE INSTRUCTIONS WERE GIVEN. WORKLOAD OF CTLR, OBTAINING RELEASES FROM BAY APCH, ISSUING TA'S TO ACFT ON RADAR, COORDINATING WITH GND CTL TAKING HDOFS FROM BAY, PREVENTS THE LCL CTLR FROM SCANNING RWYS AND, THUS, POSSIBLY PREVENTING THESE TYPES OF OCCURRENCES. THE RESPONSIBILITY PLACED UPON LIMITED RADAR CTLRS IS TOO MUCH AND SOMETHING SHOULD BE DONE TO ELIMINATE IT. SEE FAA ORDER 7110.65 SECTION 8 PARAGRAPH 7-8-2 FOR A COMPLETE DESCRIPTION OF CLASS C SVCS. COPY AVAILABLE UPON REQUEST.

SYNOPSIS PLT OF A C150 LANDED ON THE WRONG RWY AFTER BEING CLRED TO LAND. TWR CTLR BLAMES

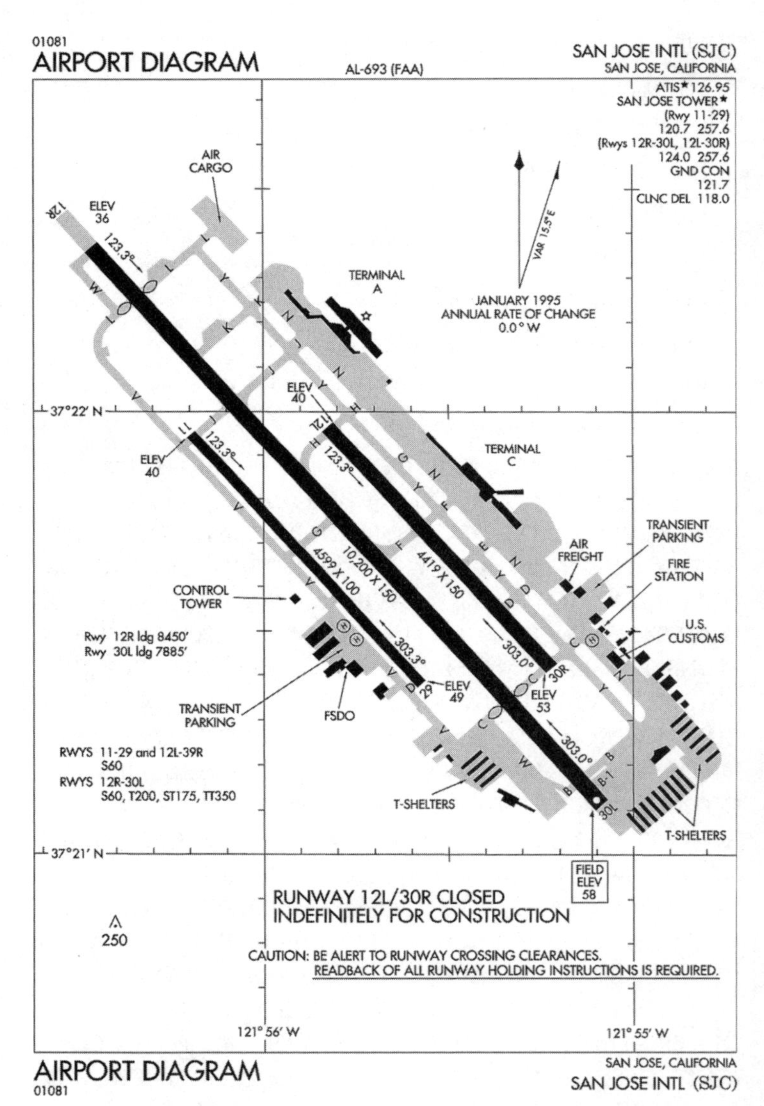

4-2 *San Jose International, SJC (this airport diagram is not suitable for navigational purposes).*

HIMSELF FOR NOT OBSERVING THE C150 LINING UP TO LAND ON THE WRONG RWY DUE TO BEING TOO BUSY WITH OTHER DUTIES.

Preparation and Remarks

From the ASRS report, it is assumed that the pilot of the Cessna 150 was not familiar with San Jose International Airport. It is also quite possible that the pilot was not using an airport diagram—perhaps leading to the confusion between runways—although 29 and 30R do not look the same when painted on the runway surface.

The controller reports that the workload was very heavy and appears to carry the blame (based on wording used).

It also appears that the reason for this report was to point out the heavy workload—on which the incident is blamed. As such, this report may have been intended as a blast directed toward the ATC system and its workload allocations.

Postincident Analysis

No reported separation loss resulted from this incident. The narrative indicates very clearly the overworked character of the controller (author of the report).

The stated controller workload consisted of obtaining releases from bay approach, giving traffic advisories, and coordinating hand-offs—thus "PREVENTS THE LCL CTLR FROM SCANNING RWYS...." This means that the controller was not visually scanning the runways and visually monitoring aircraft activity.

The pilot was lost in the air, as shown by the landing on runway 30R rather than on runway 29. The simple use of an airport diagram would have prevented this problem because the pilot at least would have been able to plan ahead and visualize the airport.

The incursion would be classified as a pilot deviation, and the runway incursion severity category would be category D because there was little or no chance of a collision. However, it meets the definition of a runway incursion because there is no mention of any loss of separation.

Several areas of concern are noted:

Problem 1. The Cessna pilot was unfamiliar with the airport and landed on the incorrect runway—totally on the far side of the airport from the intended runway.

Problem 2. Visual indicators were ignored (runway surface markings) as the pilot landed the Cessna 150.

Problem 3. The pilot may have been confused by the dimensions of the runways. The smaller runway (for which the C150 was cleared) may have appeared to be a taxiway. However, use of an airport diagram would have prevented the confusion.

Problem 4. According to the report, the controller was too busy running hither and yon to visually observe the C150 landing—a factor of work overload within the ATC system.

Lessons Learned

The lessons learned from this incident deal with pilot preparation and responsibility. Also learned is that controller workloads can affect pilots directly.

Lesson 1. Pilots using an unfamiliar airport should use an airport diagram. Use of a diagram allows the pilot to visualize the airport while approaching it. In this case, there is no doubt that this would have prevented the confusion of multiple parallel runways.

Lesson 2. Pilot observation—as in looking for runway numbers—is required. This too could have prevented the incident. Runway markings 29 and 30L, as painted on the ends of the two runways, do not appear similar.

Lesson 3. Pilot observation—by referring to the airport diagram—could have prevented this incident. The runway landed on is physically much larger (in all directions) than the intended runway. There is an obvious visual difference.

Lesson 4. A pilot cannot depend on the controller to observe and correct the pilot's mistakes.

CASE 3

Controller's Faulty Relief Instructions

ASRS accession number: 454213

Month and year: November 1999

Local time of day: 1801 to 2400

Facility: BUF, Buffalo Niagra International Airport (FIG. 4-3)

Location: Buffalo, NY

Flight conditions: IMC

Aircraft 1: DC-9

Aircraft 2: B737

Pilot of aircraft 1: Captain

Pilot of aircraft 2: Captain

Reported by: Controller

Incident description: Runway incursion

Incident consequence: Landing aircraft made a go-around.

NARRATIVE LCL AND GND CTL COMBINED. RWY 5 IN USE (APCH END DIFFICULT TO SEE BECAUSE OF POORLY MOUNTED RAMP LIGHTS, DARKNESS, RAIN AND OBSTRUCTIONS). SEVERAL MINS AFTER I ASSUMED LCL CTL AND GND

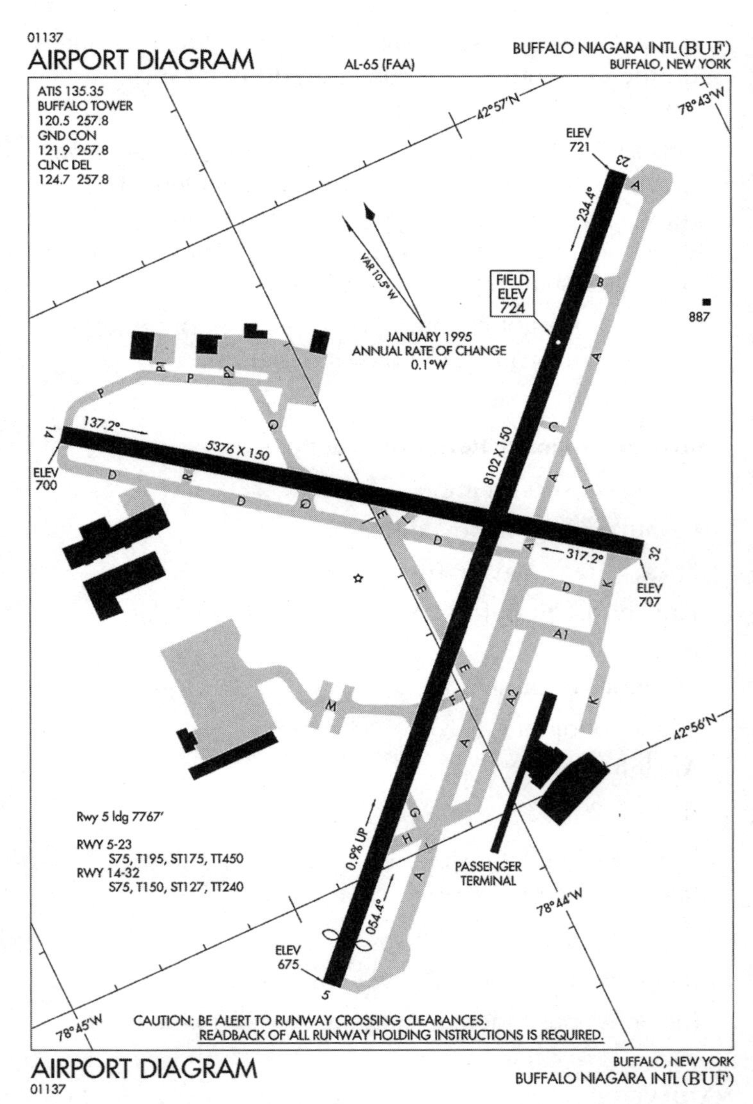

4-3 *Buffalo Niagara International, BUF (this airport diagram is not suitable for navigational purposes).*

CTL A DC9 CALLED, 'READY.' I ASSUMED THE ACFT MEANT READY TO TAXI BECAUSE ON THE POS RELIEF FROM THE OTHER CTLR NOTHING WAS MENTIONED ABOUT A DC9 TAXIING, THE ACFT IDENT WAS NOT WRITTEN DOWN ON THE SCRATCH PAD, AND THE FLT PROGRESS STRIP WAS IN THE INACTIVE BAY. I RESPONDED 'ACR X TAXI RWY 5.' ACFT RESPONDED 'ACR X ROGER.' I DID NOT HAVE ANY GND TFC BTWN DC9'S RAMP AND RWY. I LOOKED TOWARD THE RAMP AND SAW AN ACFT WHICH I ASSUMED WAS THE DC9, BUT WAS AN ACFT PARKED NEARBY. I CLRED THE B737 TO LAND (WX 400 FT OVCST). THE DC9 ASKED 'WHAT RWY WERE WE USING?' I RESPONDED RWY 5. DC9 THEN ASKED WHERE THE B737 WAS. I RESPONDED A 1½-MI FINAL. THE DC9 THEN INFORMED ME THAT HE WAS ON THE RWY. THE B737 WAS ON A SHORT FINAL AT WHICH TIME HE WAS SENT AROUND. CAUSES: POOR POS RELIEF BY PRECEDING CTLR. ASSUMPTION BY DC9 PLT THAT 'TAXI RWY 5' MEANT THAT HE COULD TAXI INTO POS. NO READ· BACK BY PLT. LIGHTING CONDITIONS. ASSUMPTION BY MYSELF THAT PARKED ACFT WAS THE TAXIING ACFT.

SYNOPSIS A DC9 AT THE BUF ARPT, HOLDING SHORT OF THE RWY, TAXIED INTO POS WHEN THE CTLR SAID, 'TAXI TO RWY 5.' TFC ON SHORT FINAL WAS SENT AROUND.

Preparation and Remarks

In general, a lack of preparation on the part of the controller(s) caused this entire incident. The previous controller did not provide the relief controller with the information necessary to perform the function of the assigned position. Specifically, the relief controller did not know where the landing aircraft (B737) was or where the taxiing aircraft (DC9) was.

Observation of parked aircraft, by the relief controller, was faulty, and he identified an incorrect aircraft.

The controller was able to prevent a loss of separation by having the landing aircraft (B737) go around.

Postincident Analysis

On the part of the controllers involved, this incident exemplified the necessity of providing the relief (oncoming) controller with the necessary information to allow an assumption of duties.

The DC-9 pilot made the incorrect assumption of having received a clearance to enter runway 5. "TAXI RWY 5" does not constitute clearance to enter runway 5. It means to go to it and hold short of it.

As a result of the DC-9 being on the runway and not observed by the controller due to weather and physical airport difficulties, the landing B737 had to make a go-around.

This runway incursion would be classified as an operational error, and the incursion category would be category B because separation decreased and there was a significant potential for collision. The landing B737 had to make a go-around.

Several problems resulted from the lack of information:

Problem 1. The oncoming controller was given no indication of the DC-9 taxiing to runway 5.

Problem 2. When the DC-9 said "READY," the controller, due to the above-mentioned lack of information, assumed this to be new aircraft movement starting from the ramp—when the DC9 actually was indicating readiness for takeoff while on or at the end of runway 5.

Problem 3. It is unclear from the report whether the DC-9 had lined up on the end of runway 5 with permission from the previous controller and then said "READY." The controller is to be lauded for hastily instructing the B737 to make a go-around, thereby preventing a disaster.

Problem 4. The DC-9 flight crew should have been more vigilant in monitoring what was happening around them—specifically, the landing B737. This would have been relatively easy because there was very little traffic at the time of the incident.

Problem 5. There is no indication that the landing flight crew ever saw the DC-9 sitting on the runway.

Lessons Learned

The lessons learned from this incident are relative to the operation in the cab of the tower and to flight crew monitoring radio transmissions. The incident shows clearly how runway incursion incidents can be caused by pilots or by controllers—or by a combination of efforts.

Lesson 1. The relief controller must obtain the necessary information from the previous controller to perform the assigned duties safely. Specifically, the relief controller needs to know what aircraft are in operation at the airport (ground movement) and in the vicinity of the airport (on approach).

Lesson 2. Not mentioned in the report is the radio traffic, which should have been monitored by the DC-9 flight crew. This traffic would have been an earlier landing clearance given to the B737, possibly from the previous controller, indicating that the end of runway 5 might be in conflict. The prudent flight crew would have questioned the controller and not entered the runway until specifically cleared to do so.

CASE 4

Heads Up Until in the Chocks

ASRS accession number: 457053

Month and year: December 1999

Local time of day: 0601 to 1200

Facility: BWI, Baltimore-Washington International

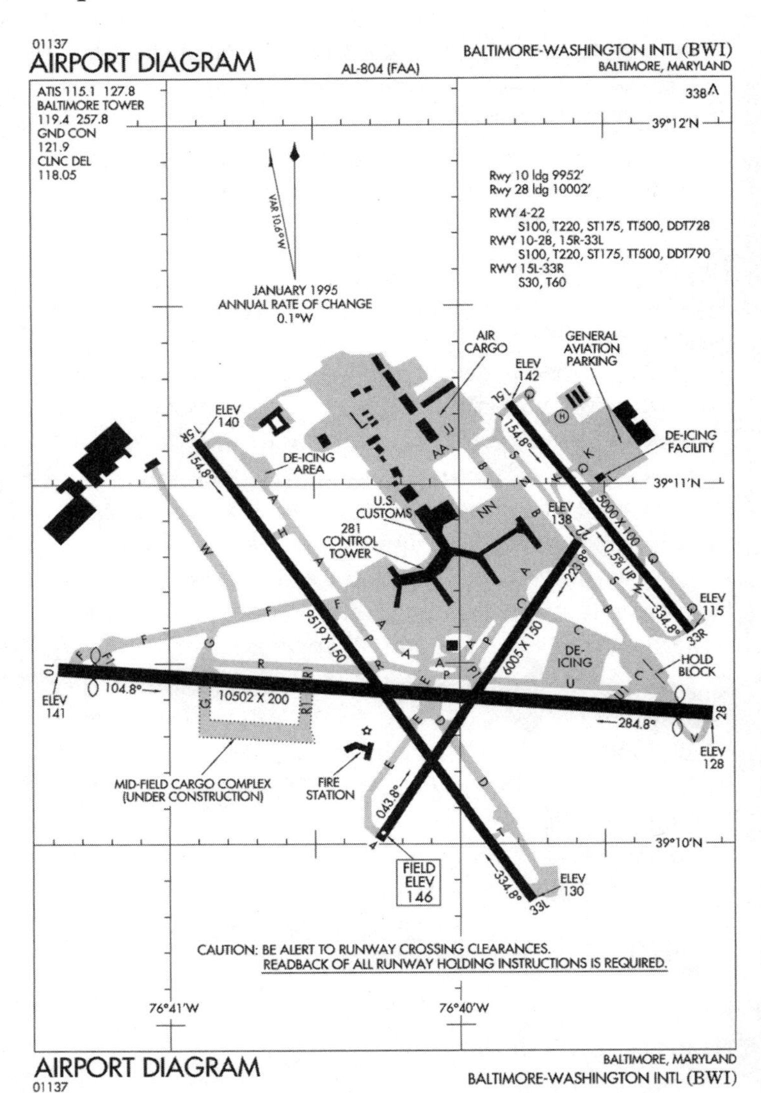

4-4 *Baltimore-Washington International, BWI (this airport diagram is not suitable for navigational purposes).*

Airport (FIG. 4-4)

Location: Baltimore, MD

Flight conditions: VMC

Aircraft 1: B727

Pilot of aircraft 1: Captain; first officer; second officer, 6000 hours

Reported by: Pilot

Incident description: Runway incursion

Incident consequence: None

NARRATIVE WE HAD CLRED THE RWY AND WERE GIVEN CLRNC TO TAXI TO PARKING. LANDED RWY 33L AND WERE TAXIING TO CARGO SPOT AB—JUST BEYOND TXWY J ON TXWY A. WE WERE TAXIING ON TXWY P AND MISSED TURN TO TXWY JUST PRIOR TO RWY 4/22. WE WERE ON RWY 4/22 BEFORE WE REALIZED IT. NO OTHER ACFT WERE INVOLVED AND GND CTL SAID NOTHING. SUSPECT THE TURN FROM TXWY P TO TXWY P JUST BEFORE RWY 4/22 IS NOT MARKED VERY WELL. ALSO DO NOT REMEMBER LIGHTS OR HOLD SHORT LINES SO THEY MAY HAVE BEEN OUT OR FADED. CORRECTIVE ACTION SUGGESTED: LOOK AT THIS INTXN AT NIGHT FROM COCKPIT LEVEL. REBRIEF PLTS THAT THESE EVENTS CAN AND DO HAPPEN. KEEP HEADS UP TILL IN CHOCKS.

SYNOPSIS FLC OF A B727 INADVERTENTLY TAXIED ONTO RWY DURING TAXI IN AFTER LNDG DUE TO THE DAWN LIGHTING CONDITIONS AND NOT NOTICING PROMINENT SIGNS AND HOLD-SHORT MARKINGS.

Preparation and Remarks

This incident involved a professional flight crew, and it is assumed that they were working with a proper airport diagram. Although causing no separation loss, their aircraft did enter runway 4/22.

The flight crew claims that signage and surface markings were not observed, possibly due to poor dawn lighting conditions.

Note, however, the comment in the synopsis saying, "NOT NOTICING PROMINENT SIGNS AND HOLD SHORT MARKINGS."

Postincident Analysis

There appears to have been a lack of attention to the runway diagram for reference purposes and recognition that there may have been visual difficulty caused by dawn lighting conditions.

The ASRS report first states, "SUSPECT THE TURN FROM TXWY P TO TXWY P JUST BEFORE RWY 4/22 IS NOT MARKED VERY WELL. ALSO DO NOT REMEMBER LIGHTS OR HOLD-SHORT LINES SO THEY MAY HAVE BEEN OUT OR FADED." Then it later indicates that there were prominent signs and hold-short markings. Unfortunately, excuses aside, an incursion did happen—and the tower never noticed.

The classification of this incursion would be pilot deviation, and the runway incursion severity category would be category D because there was little or no chance of a collision. However, it meets the definition of a runway incursion even though no other aircraft were involved.

The following problems are noted regarding this incident:

Problem 1. Reference to the runway diagram shows that taxiway P turns to the left prior to runway 4/22 and is also intersected by a taxiway going to the runway P1. This intersection was missed, and the aircraft entered runway 4/22.

Problem 2. The flight crew indicated that it saw no lights, signs, or markings indicating the end of the taxiway. The synopsis mentions "NOT NOTICING PROMINENT SIGNS AND HOLD SHORT MARKINGS," leaving you with two differing points of view. There also was no mention of other signage.

Problem 3. The controller missed the incursion.

Lessons Learned

From this incident, which probably would not have occurred had the flight crew been more vigilant in airport diagram use, we see how differing opinions sometimes are used to cover for mistakes. The signs and lights were either out or faded, or they were on and clear. Either way, they were there and should have been seen.

Lesson 1. Use airport diagrams, referring to them often while watching for surface markings, lighting, and signs where you would expect to see them, based on the airport diagram.

Lesson 2. The runway incursion happened, regardless of the excuses offered. The flight crew is responsible. According to FAR 91.3 (a), "The pilot in command of an aircraft is directly responsible for, and is the final authority to, the operation of that aircraft." The FAR does not provide for mitigating circumstances such as sun in the eyes or poor signage.

Lesson 3. As in the ASRS report, "KEEP HEADS UP TILL IN CHOCKS." The airplane is still flying until it is parked. This means that the flight crew is still flying also (the pilot in command is responsible) and must remain vigilant at all times.

CASE 5

A Captain's Decision

ASRS accession number: 405084

Month and year: June 1998

Local time of day: 1801 to 2400

Facility: PIT, Pittsburgh International Airport (FIG. 4-5)

Location: Pittsburgh, PA

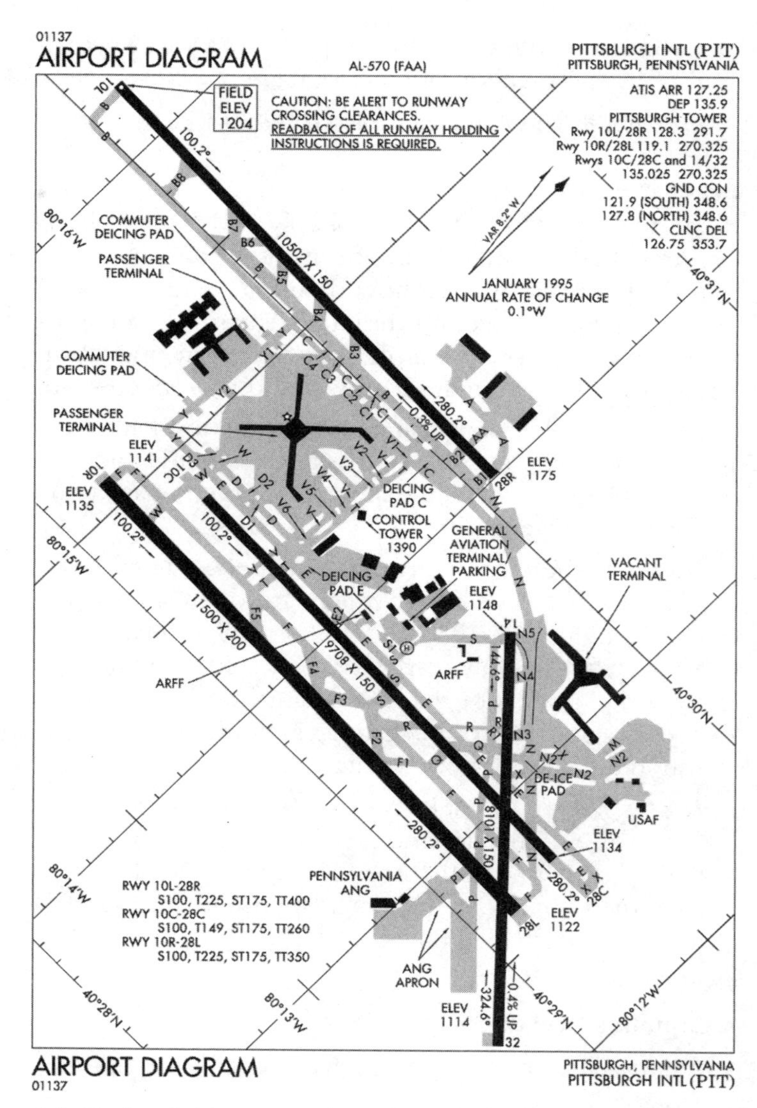

4-5 *Pittsburgh International, PIT (this airport diagram is not suitable for navigational purposes).*

Flight conditions: VMC

Aircraft 1: Fokker 100

Pilot of aircraft 1: Captain, 13,000 hours; first officer

Reported by: Captain

Incident description: Refused LAHSO clearance

Incident consequence: None

NARRATIVE PIT TWR CLRED US TO LAND AND HOLD SHORT OF RWY 14/32. I REFUSED THIS CLRNC BY STATING WE WERE UNABLE TO HOLD SHORT OF RWY 14/32. WE WERE THEN CLRED TO LAND AND MANAGED TO TURN OFF EARLY. WHILE TURNING OFF, WE WERE TOLD TO HURRY UP AS OTHER ACFT WERE WAITING AND TO CALL THE TWR ON THE LANDLINE. AFTER CALLING TWR SUPVR HE WANTED TO KNOW WHY WE REFUSED THE LAHSO. I STATED THAT IT WAS A CLRNC WITH THE CAPT'S DISCRETION. HE WAS OF THE IMPRESSION THAT A LAHSO CLRNC WAS MANDATORY. OUR FLT RTE MANUAL CLRLY STATES 'THE ACCEPTANCE OF A LAHSO CLRNC IS AT THE CAPT'S DISCRETION.' THIS IS OBVIOUSLY THE CASE DUE TO NUMEROUS POSSIBLE FLT LIMITATIONS, IE, MELS OR ACFT PERFORMANCE, OR THE FACT THAT THE APCH WOULD BE UNCOMFORTABLE FROM A SAFETY STANDPOINT. TWR PERSONNEL SHOULD BE AWARE OF THESE FUNDAMENTAL POLICIES AND IF UNSURE SHOULD LOOK THEM UP PRIOR TO CONTACTING PLTS. THIS SUPVR WAS MORE CONCERNED WITH THE SPD OF DEP OPS THAN SAFETY. ADDITIONALLY, I FOUND THE TABLES DEPICTING THE AUTH FOR LAHSO OPS FOR VARIOUS RWYS DIFFICULT TO QUICKLY INTERP. IN ADDITION THE PIT 3 LAHSO INFO IS IN ERROR AND IN CONFLICT WITH NOTAM FOR RWY 10C LAHSO. CALLBACK CONVERSATION WITH RPTR REVEALED THE FOLLOWING INFO: THE CAPT SAID THAT HIS F100 PERFORMANCE MANUAL CONTAINED A PROHIBITION AGAINST LAHSO'S DURING WET CONDITIONS ON RWY 10C. HE ADDED THAT HE HAD OPERATED IN AND OUT OF PIT SEVERAL TIMES THAT DAY AND IT HAD RAINED. HE HAD

ASSUMED THAT THE RWY WAS STILL WET SINCE HE COULD NOT SEE IT AT NIGHT. HE FOUND THAT THE PERFORMANCE DATA WAS MORE DIFFICULT TO USE THAN IT SHOULD HAVE BEEN TO DETERMINE IF HE COULD LAND AND HOLD SHORT. HE ALSO FELT THAT THE HOLD SHORT DATA WAS INCORRECT IN HIS COMMERCIAL CHART MANUAL. HE HAS FORWARDED THIS INFO TO HIS COMPANY. THE RPTR SAID THAT HE CALLED THE TWR AT THEIR REQUEST AND THE SUPVR WAS RATHER UNPLEASANT AND THREATENING TOWARD HIM. HE DETERMINED THAT THE SUPVR WAS UNDER THE IMPRESSION THAT HE WAS REQUIRED TO HOLD SHORT IF ORDERED ON LNDG. THE RPTR TOLD HIM THAT THIS TYPE OF CLRNC WOULD BE ACCEPTED AT PLT'S DISCRETION.

SYNOPSIS AN ACR FK10 FLC RPTS THAT A TWR CTLR ATTEMPTED TO CONVINCE THEM TO ACCEPT A LAHSO CLRNC ON A RWY THAT THE FLC CONSIDERED WET. THE FLC BELIEVES THAT THE CTLR THOUGHT THAT HIS HOLD SHORT CLRNC WAS MANDATORY.

Preparation and Remarks

The flight crew was prepared for arrival at Pittsburgh with all necessary information at hand. It is apparent from the text of the narrative that air traffic conditions at the time of the incident were heavy.

LAHSO (land and hold short operations) are a continuing problem for flight crews and controllers. LAHSO is the result of airport overcrowding. The intent of LAHSO is to allow more aircraft to operate from the airport than otherwise would be possible. Although issued by ATC, the clearances for LAHSO are always at the discretion of the aircraft commander.

Several factors can contribute to the decision to accept or not to accept a LAHSO clearance. Some include weather conditions, runway conditions, aircraft limitations, and company policies—to name a few. Always, safety is the underlying factor.

Postincident Analysis

There was no separation loss involved with this incident and no incursion as such. The incident resulted from a decision by the aircraft commander to refuse a LAHSO clearance (which is the commander's choice). LAHSO is a tool allowing arrivals on one runway and departures on another—even though the runways intersect. The object is to have the landing aircraft hold short of the active departure runway and not interfere with operations on that runway.

Several problems are noted from this specific incident:

Problem 1. The captain refused the LAHSO clearance—which is clearly the captain's choice. There is no argument with the choice made.

Problem 2. After landing, the aircraft was told to "HURRY UP AS OTHER ACFT WERE WAITING" and for the pilot to call tower on the telephone. This sounds like harassment of the flight crew for not accepting the LAHSO clearance.

Problem 3. The ATC supervisor did not understand that a LAHSO was not mandatory, as shown by the statement, "AFTER CALLING TWR SUPVR, HE WANTED TO KNOW WHY WE REFUSED THE LAHSO. I STATED THAT IT WAS A CLRNC WITH THE CAPT'S DISCRETION. HE WAS OF THE IMPRESSION THAT A LAHSO CLRNC WAS MANDATORY."

Lessons Learned

LAHSO has presented many problems to pilots. The most noteworthy being a "hurry up and get out of the way" attitude from ATC. Many pilots are concerned that LAHSO is not safe because it removes a portion of the available runway during the landing.

LAHSO is viewed by airport operators as a tool to get more aircraft in and out of the airport (with more and more passengers) than would be allowed safely with normal aircraft separation and runway use.

LAHSO typically is viewed by pilots as a dollar problem for airport operators and airlines—only interested in moving the most possible passengers in the least time and space possible. Many pilots do not feel that LAHSO operations are safe, and none feel that LAHSO is safe under all conditions.

Several lessons can be learned from this incident:

Lesson 1. The flight commander has the final say on the acceptance of a LAHSO clearance. It may be accepted or refused for various physical reasons, such as aircraft performance, runway limitations, weather, etc. It also may be refused because the pilot is not comfortable with the clearance—no later justification is needed. Refusal is a judgment call.

Lesson 2. It is not the place of overzealous controllers to browbeat flight crews into accepting LAHSO clearances. Flight safety is the ultimate responsibility of the aircraft's commander.

CASE 6

Lost Communications

ASRS accession number: 460325

Month and year: January 2000

Local time of day: 1201 to 1800

Facility: ILG, Wilmington–New Castle County Airport (FIG. 4-6)

Location: Wilmington, DE

Flight conditions: VMC

Aircraft 1: PA-28

Pilot of aircraft 1: Single pilot, 71.5 hours

Reported by: Pilot

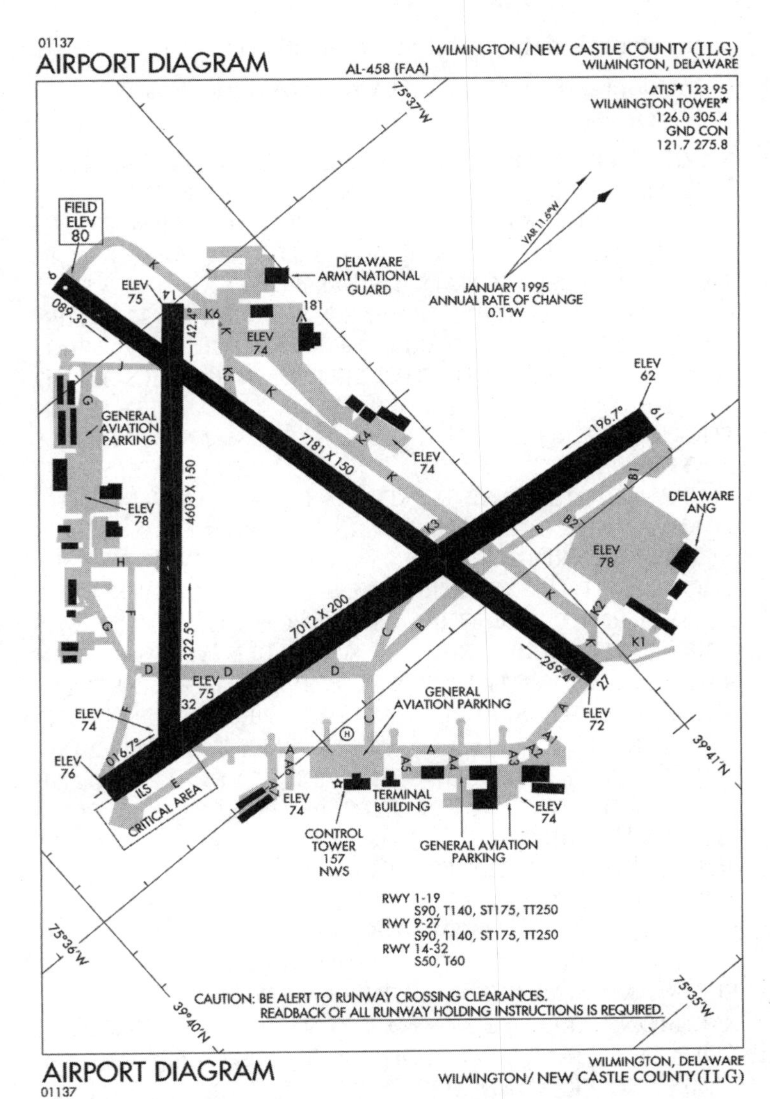

4-6 *Wilmington/New Castle Airport, ILG (this airport diagram is not suitable for navigational purposes).*

Incident description: Runway incursion

Incident consequence: FAA reviewed incident with pilot.

NARRATIVE I AM A STUDENT PLT. I FLEW TO THE PRACTICE AREA FOR MANEUVERS, AND THEN I HEADED BACK TO THE ARPT WITH THE INTENTION OF PRACTICING TOUCH-AND-GO LNDGS. I RADIOED THE ARPT ON CTL TWR FREQ 126.00. I HAD JUST FINISHED MY DSCNT TO PATTERN ALT ON THE L DOWNWIND FOR RWY 27, WHEN I SUDDENLY ENCOUNTERED STRONG TURB. MY HEAD SLAMMED UP AGAINST THE CEILING, WHICH KNOCKED MY HEADSET PARTLY OFF. THIS OCCURRED EVEN THOUGH I HAD MY SHOULDER HARNESS AND LAP BELT FASTENED LOW AND TIGHT. I RECOVERED QUICKLY AND RESTORED THE PLANE TO STRAIGHT AND LEVEL FLT. THE TURB SUBSIDED. AT THAT POINT, I NOTICED AN EMER ALARM SOUNDING IN MY EARS. I IMMEDIATELY THOUGHT THAT MY AIRPLANE'S ELT MUST HAVE BEEN ACTIVATED BY THE TURB. ALTHOUGH THE ALARM WAS LOUD AND INTERFERING, I WAS ABLE TO MAINTAIN RADIO COMS WITH TWR. I RECEIVED LNDG INSTRUCTIONS. TWR COMS WERE BUSY, AND I DECIDED I SHOULD NOT TAKE TIME TO MAKE ANY COMMENT CONCERNING THE ELT. AFTER TOUCHDOWN, I EXITED RWY 27 AT THE VERY FIRST INTXN. I STOPPED THE AIRPLANE AND COMPLETED MY AFTER LNDG CHKLIST. THE ELT ALARM WAS SOUNDING MUCH LOUDER IN MY EARS AFTER I GOT ON THE GND, AND I COULD NOT SEEM TO HEAR ANY RADIO COMS. I DECIDED TO SWITCH TO GND CTL FREQ 121.70 HOPING THAT I COULD GET TAXI INSTRUCTIONS. THE ELT ALARM WAS EVEN LOUDER IN MY EARS. I MADE 2 XMISSIONS ATTEMPTING TO GET TAXI INSTRUCTIONS. I COULDN'T HEAR ANY COHERENT RESPONSES TO MY COMS OTHER THAN LOW STATIC RUMBLING AND AN OCCASIONAL BROKEN WORD OR TWO IN THE BACKGND. I DECIDED THAT THE SIT WAS UNSAFE, AND THAT I SHOULD TAKE SOME ACTION TO CLR THE AREA SAFELY. I

BEGAN TO TAXI WHEN I HEARD SOME STATIC RUMBLING AND BROKEN WORDS UNDER THE ELT ALARM SOUND. I STOPPED THE ACFT AND XMITTED AGAIN ON 121.70, SAYING I WAS UNABLE TO COPY ANY VOICE XMISSION. I REALIZE NOW THAT I SHOULD NOT HAVE TAXIED WITHOUT CLRLY UNDERSTOOD INSTRUCTIONS. I ASSUMED THAT ANYONE ELSE WAS HEARING THE SAME LOUD ALARM I WAS HEARING, AND THAT THEY UNDERSTOOD MY AIRPLANE WAS THE SOURCE. I LOOKED UP AT THE TWR WINDOWS FOR A FEW MOMENTS, AND DECIDED ONCE AGAIN THAT THIS WAS AN EMER, THAT I SHOULD CLR THE AREA SAFELY AS I COULD. EVEN THOUGH I HAD BEEN TRAINED TO LOOK FOR LIGHT SIGNALS IN THE EVENT OF RADIO COMS FAILURE, I MISTAKENLY FAILED TO EQUATE THIS SIT WITH A 'RADIO OUT' SIT. I DIDN'T SEE ANY LIGHT SIGNALS, BUT I HAVE TO CONFESS THAT I ALSO FORGOT TO LOOK FOR THEM. I WAS CONCERNED THAT MY PRESENCE ON THE ARPT WITH AN ACTIVE ELT CONSTITUTED A DANGER TO OTHERS. THE MAIN THOUGHT IN MY MIND WAS THAT I SHOULD TAXI BACK TO THE FBO SO SOMEONE COULD DEACTIVATE THE XMITTER. I TAXIED TO THE HOLD SHORT LINE AT RWY 1/19 WHILE I LOOKED FOR TFC. I SAW NO TFC, SO I TAXIED ACROSS RWY 1/19 AND CONTINUED TOWARDS RWY 14/32. AT THAT POINT, AN ARPT SAFETY VEHICLE SUDDENLY APPEARED ON MY L WINGTIP WITH LIGHTS FLASHING. I ALSO HEARD THE FIRST CLR RADIO XMISSION SINCE I HAD LANDED, WITH A VOICE SAYING CLRLY ABOVE THE SOUND OF THE ELT ALARM, 'STOP THE AIRPLANE!' THE SAFETY VEHICLE PULLED IN FRONT OF ME, WAVING WITH A GESTURE THAT SAID 'FOLLOW ME.' I FOLLOWED THE VEHICLE ACROSS RWY 14/32. AT POINT, THE VEHICLE GESTURED FOR ME TO CONTINUE TO THE FBO. I PARKED THE AIRPLANE, AND I WAS IMMEDIATELY JOINED BY A MECH, AS WELL AS THE ARPT SAFETY OFFICER. THE MECH VERIFIED THE ELT HAD BEEN ACTIVATED. THE SAFETY OFFICER ASKED ME TO DESCRIBE WHAT HAD HAPPENED. HE THEN ASKED ME TO GIVE TWR A

PHONE CALL. MY FLT INSTRUCTOR SAID HE COULD SEE THE ELT LYING ON THE FLOOR. BACK IN THE FBO, MY FLT INSTRUCTOR CALLED TWR AND TALKED WITH THE CTLR. MY FLT INSTRUCTOR THEN SAID I HAD COMMITTED A RWY INCURSION AND GAVE ME THE PHONE. I WAS VERY SURPRISED TO BE TOLD THAT I HAD COMMITTED THIS VIOLATION. THE CTLR GAVE ME A STERN LECTURE. HE TOLD ME THAT I SHOULD HAVE TUNED MY RADIO BACK TO TWR FREQ 126.00, WHERE I MIGHT HAVE BEEN SUCCESSFUL IN GETTING TAXI INSTRUCTIONS, OR I SHOULD HAVE WAITED FOR LIGHT SIGNALS, OR FOR A SAFETY VEHICLE TO COME TO MY LOCATION ON THE TXWY. I SPENT MORE TIME TALKING WITH MY FLT INSTRUCTOR ABOUT WHAT I SHOULD HAVE DONE TO DEAL WITH THE SIT SAFELY AND PROPERLY. I SHOULD HAVE REALIZED THAT THIS SIT CALLED FOR ME TO FOLLOW THE STANDARD 'RADIO OUT' PROCS I HAD BEEN TAUGHT FOR COMMUNICATING VIA LIGHT SIGNALS. I NOW UNDERSTAND THAT A SMALL AIRPLANE PARKED ON A WIDE TXWY DOES NOT CONSTITUTE NEARLY SO GREAT A DANGER AS THE SAME SMALL AIRPLANE IF IT TAXIES WITHOUT PROPER COMS AND CLRNC.

SYNOPSIS PA28 STUDENT PLT ENCOUNTERS TURB IN TFC PATTERN ACTIVATING ONBOARD ELT. AFTER LNDG, WITH THE INTERFERING ELT ALARM, PLT CROSSES RWY WITHOUT CLRNC. ARPT VEHICLE INTERVENES, GUIDING ACFT TO FBO WHERE INCIDENT IS REVIEWED WITH SAFETY OFFICER AND INSTRUCTOR.

Preparation and Remarks

The student pilot making this ASRS report indicated that training was received for no radio procedures; however, during the incident, the training was forgotten (not applied). There was no problem in being familiar with the airport. The severe turbulence encountered in the traffic pattern could not be planned for.

Postincident Analysis

No loss of aircraft separation occurred as a result of this incident. The incident does, however, show how training can be forgotten when it is needed most. Had the procedure been remembered for lost communications, this incident would not have happened.

The incursion is classified as a pilot deviation, and the runway incursion severity category would be category D because there was little or no chance of a collision. However, it meets the definition of a runway incursion.

The tower was very aware of a problem on the runway and was observing it. Sending an airport safety vehicle to the aircraft prevented further incursion.

Problem 1. The aircraft encountered severe turbulence, which cannot be planned for.

Problem 2. As a result of the turbulence, the emergency locator transmitter (ELT) sounded and blocked reception of controller instructions. Lost communications can be planned for.

Lessons Learned

The primary lesson learned in this incident regards what to do when you lose radio communications with the tower. The pilot admitted to having been trained for such a loss of communications, but forgot what to do.

Lesson 1. Have a plan for operating after the lose of radio communications. See AIM 4-2-13 for further information.

Lesson 2. Regardless of the circumstances, you are always safer stopping on a taxiway and waiting for help than continuing to taxi and crossing one or more runways.

Lesson 3. Keep the seat belt/shoulder harness tight. You never know when you may be affected by severe

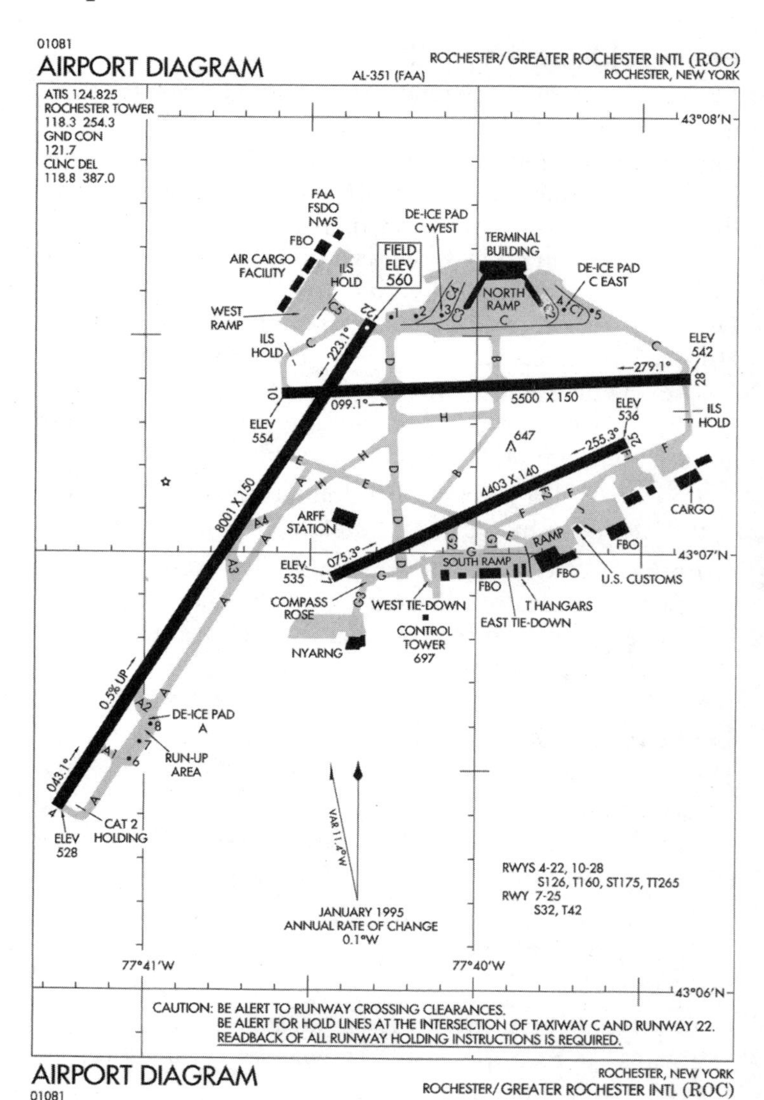

4-7 *Rochester International, ROC (this airport diagram is not suitable for navigational purposes).*

turbulence, causing an unusual aircraft attitude or causing you to fall from the seat.

CASE 7

Alert Flight Crews

ASRS accession number: 446246

Month and year: August 1999

Local time of day: 1201 to 1800

Facility: ROC, Rochester–Greater Rochester International Airport (FIG. 4-7)

Location: Rochester, NY

Flight conditions: VMC

Aircraft 1: B727

Aircraft 2: Astra Jet

Pilot of aircraft 1: Captain; first officer, 2600 hours

Pilot of aircraft 2: Captain

Reported by: Pilot aircraft 1

Incident description: Near runway incursion

Incident consequence: Critical ground conflict

NARRATIVE WHILE INBOUND TO ROC, WE WERE BEING VECTORED FOR RWY 22 VISUAL APCH WHEN WE (B727) REQUESTED A VISUAL APCH TO RWY 4. APCH TOLD US TO STAND BY WHILE THEY CHKED WITH THE TWR. THEY CAME BACK AND GAVE US A VECTOR TO RWY 4 AND CLRED US FOR THE VISUAL TO RWY 4 AND TOLD US TO CONTACT THE TWR. UPON CONTACTING THE TWR THEY CALLED TFC ON A 4 MI FINAL TO RWY 28 AND CLRED US TO LAND. AFTER TOUCHDOWN AND DEPLOYMENT OF THRUST REVERSERS, WE NOTICED THE ASTRA JET ROLLING OUT ON RWY 28 AND NEARING THE INTXN OF RWY 4. WE APPLIED MORE PRESSURE TO THE BRAKES AND CAME TO A STOP SHORT OF THE RWY 4/28 INTXN. UPON STOPPING, THE ASTRA JET CALLED THE TWR AND

ASKED THEM IF WE HAD BEEN INSTRUCTED TO HOLD SHORT OF THE INTXN AND THEIR RESPONSE WAS 'NEGATIVE.' HE THEN ASKED IF THEY (ASTRA JET) WERE SUPPOSED TO HOLD SHORT OF THE INTXN AND AGAIN THE TWR'S RESPONSE WAS 'NEGATIVE.' THE ASTRA JET THEN INFORMED THE TWR THAT THEY HAD INTENDED TO ROLL THROUGH THE INTXN TO THE END OF THE RWY. THE TWR THEN TOLD THE ASTRA JET TO HOLD THEIR POS WHILE THEY TOLD US TO TAXI TO PARKING. I BELIEVE THAT THE VIGILANCE OF BOTH FLCS IS THE FACTOR THAT PREVENTED THIS FROM BECOMING AN ACCIDENT WHERE MANY PEOPLE COULD HAVE BEEN INJURED OR KILLED.

SYNOPSIS B727-200 CONFLICTS WITH ANOTHER ACFT LNDG ON INTERSECTING RWY WITH NO TWR LAHSO OR OTHER COMMENT AT ROC.

Preparation and Remarks

The flight crews both appear to have been prepared and were adhering to ATC instructions. There was no apparent confusion about positions or instructions.

The controller was not properly aware of the two aircraft landing on intersecting runways.

Postincident Analysis

This incident could have been very serious, but due to the vigilance of the flight crews, a collision was avoided. From the narrative of this report it is apparent that had the flight crews not been heads up, there would have a major incident with resulting loss of life and property.

The problem was very simple: two aircraft landing at the same time on intersecting runways. The flight crew of the B727 saw the other aircraft coming down runway 28 and took evasive action: "WE APPLIED MORE PRESSURE TO THE BRAKES AND CAME TO A STOP SHORT OF THE RWY 4/28 INTXN."

When the pilot of the Astra Jet "CALLED THE TWR AND ASKED THEM IF WE HAD BEEN INSTRUCTED TO HOLD SHORT OF THE INTXN AND THEIR RESPONSE WAS 'NEGATIVE.' HE THEN ASKED IF THEY (ASTRA JET) WERE SUPPOSED TO HOLD SHORT OF THE INTXN, AND AGAIN THE TWR'S RESPONSE WAS 'NEGATIVE.'" From this information, it is clear that the controller was in error and not cognizant of two aircraft landing on intersecting runways. At this time, the controller instructed the Astra Jet to hold and let the B727 taxi.

No actual incursion was involved with this event, but had there been, it would have been classified as operational error, and the runway incursion severity category would have either been category A because a radical evasive action would have been required to avoid collision or an accident involving the collision of both aircraft.

Several problems were noted in studying this incident:

Problem 1. The controller had not told either aircraft to hold short of the runway intersection where runways 4/22 and 10/28 cross.

Problem 2. When the controller was queried about holding short, the only reply was "NEGATIVE," indicating that neither aircraft had been told to hold short of anything and that both aircraft had been cleared to land on the intersecting runways.

Problem 3. Only after the problem at the runway intersection was brought to the controller's attention was a hold order issued.

Lessons Learned

The lessons learned from this incident are, for the flight crews, quite simple:

Lesson 1. Be visually vigilant at all times—looking for other air traffic. In this incident, vigilance prevented a collision.

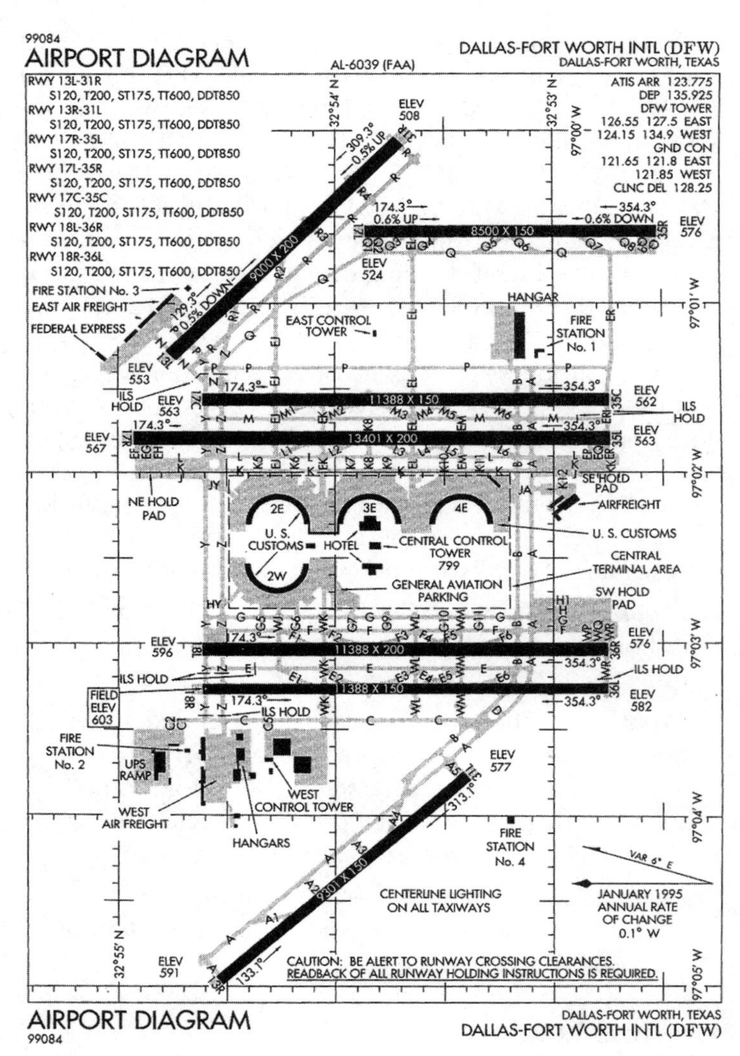

4-8 *Dallas-Fort Worth International, DFW (this airport diagram is not suitable for navigational purposes).*

Lesson 2. Monitor all airport radio transmissions, giving you the overall picture of airport activity. The B727 flight crew was aware that there was an aircraft on a 4-mile final to runway 28.

The lesson learned for the air traffic controller is also simple:

Lesson 1. Be alert! Watch your aircraft and visualize where they are going to be as times progresses—in this case, both at a runway intersection at the same time!

CASE 8

Missing Hold-Short Instruction

ASRS accession number: 444180

Month and year: July 1999

Local time of day: 0601 to 1200

Facility: DFW, Dallas–Fort Worth International Airport (FIG. 4-8)

Location: Dallas–Fort Worth, TX

Flight conditions: VMC

Aircraft 1: MD-80

Pilot of aircraft 1: Captain; first officer

Reported by: Pilot of aircraft 1

Incident description: Missed hold-short instruction

Incident consequence: None

NARRATIVE APCH SWITCHED US TO TWR, AND THEY CLRED US TO LAND, WHICH WE ACKNOWLEDGED. THE CAPT NOTICED XING TFC AT THE OPPOSITE END OF RWY AND REMARKED THAT IT WAS STRANGE THAT THE TWR HAD NOT SAID TO LAND AND HOLD SHORT FOR XING TFC. WE LANDED AND EXITED ON THE HIGH SPD. WE BEGAN TAXIING S WHEN WE HEARD THE TWR CALL. WE ACKNOWLEDGED AND THEY ASKED IF WE HEARD

THAT XMISSION, AND THAT WE HAD BEEN CLRED TO LAND AND HOLD SHORT, AND WE NEVER ACKNOWLEDGED. WE REPLIED WE HAD RECEIVED AND ACKNOWLEDGED LNDG CLRNC, BUT NOT THE HOLD-SHORT INSTRUCTIONS.

SYNOPSIS AN MD80 FLC DID NOT HEAR, OR READ BACK, THE HOLD SHORT PORTION OF THE LNDG CLRNC AT DFW.

Preparation and Remarks

The MD80 flight crew was prepared for the landing and cleared for same. There is no indication of any confusion about the airport layout or problems noted about radio communications in the report.

Postincident Analysis

There was no loss of separation resulting from this incident. The MD80 landed and exited the runway on the high-speed taxiway after the flight crew observed crossing traffic ahead at the end of the landing runway.

It was at the time of the flight crew's sighting crossing traffic on the far end of the landing runway that the captain remarked "THAT IT WAS STRANGE THAT THE TWR HAD NOT SAID TO LAND AND HOLD SHORT FOR XING TFC."

Just after that, ATC called and "ASKED IF WE HEARD THAT XMISSION, AND THAT WE HAD BEEN CLRED TO LAND AND HOLD SHORT, AND WE NEVER ACKNOWLEDGED."

The flight crew acknowledged that it had received the landing clearance, but it had not received the hold-short instruction. Note that a read-back is required when a hold-short instruction is issued. The flight crew did not receive the instruction and therefore did not read back the instruction, and the controller did not challenge the aircraft on the missing read-back.

In reading the report, notice that only after the MD80 has exited the runway onto the high-speed taxiway did the controller mention a hold-short instruction. If the controller had queried the aircraft for a read-back, the error would have been realized and corrected immediately. This is why read-backs are necessary—part of the checks and balances.

The incident is not actually a runway incursion, but it is an example of placing blame that does not seem equitable. The classification would have been pilot deviation, and the runway incursion severity category would have been category C because the aircraft did use the high-speed taxiway to clear the runway quickly.

A single important problem can be noted from this report:

Problem 1. The controller should have challenged the MD80 when the hold-short instruction was issued and not read back.

Lessons Learned

Although only a single problem was noted, there are important lessons for the flight crew and controller from this incident:

Lesson 1. Heads up! Once again, an observant flight crew noticed traffic crossing the far end of the active runway and took immediate action (exited onto the high-speed taxiway).

Lesson 2. The controller was watching the landing traffic visually and observed that there was no holding short when the MD80 exited onto the high-speed taxiway.

Lesson 3. The controller should have queried the MD80 about the hold-short instruction when the flight crew failed to read it back.

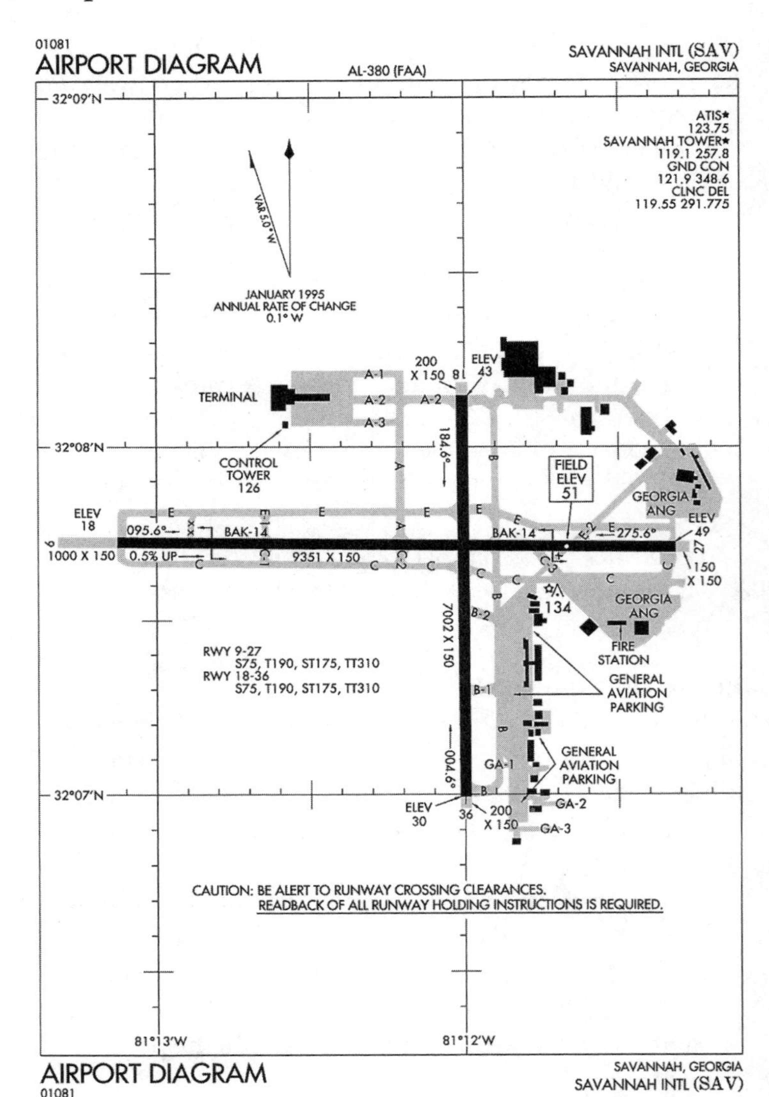

4-9 *Savannah International, SAV (this airport diagram is not suitable for navigational purposes).*

CASE 9

The Frazzled Student Pilot

ASRS accession number: 409680

Month and year: July 1998

Local time of day: 0601 to 1200

Facility: SAV, Savannah International Airport (FIG. 4-9)

Location: Savannah, GA

Flight conditions: Marginal

Aircraft 1: PA-34 Piper Seneca

Pilot of aircraft 1: Pilot and instructor

Reported by: Pilot

Incident description: Failure to follow landing instruction

Incident consequence: Controller intervened.

NARRATIVE DURING MY SECOND FLT IN THE PIPER SENECA PA34-200 I WAS RECEIVING DUAL INSTRUCTION. FOLLOWING MANEUVERS IN THE LCL TRAINING AREA I CALLED APCH CTL (SAVANNAH) AND STATED THAT WE WOULD LIKE TO RETURN FOR TOUCH AND GOES. MY INSTRUCTOR AND I HAD DIFFICULTY SEEING THE ARPT AND WE WERE GIVEN VECTORS TO FINAL. THE TWR FIRST TOLD US TO LAND (RWY 9) AND HOLD SHORT OF RWY 36. WE SAW THE ARPT AT 4.5 MI AND FLEW THE VISUAL APCH. THE TWR CANCELED OUR HOLD SHORT OF RWY 36 AND ADVISED US THAT THE FULL LENGTH WAS AVAILABLE. WE LANDED AND DID A TOUCH AND GO. THE TWR CTLR ADVISED US IF WE WANTED TO DO A TOUCH AND GO WE SHOULD HAVE ADVISED THEM. WE WERE APPARENTLY TOLD TO ENTER L TFC. MY INSTRUCTOR QUESTIONED ME WHEN WE TURNED INTO R TFC, BUT I ASSURED HIM WE WERE ADVISED R TFC. THE TWR CORRECTED US BUT TOLD US TO CONTINUE. MY LACK OF EXPERIENCE IN THIS TYPE OF ACFT, OUR

CONCERN WITH THE REDUCED VISIBILITY, AND MY RELIANCE ON OUR INTENTIONS BEING RELAYED TO THE TWR FROM APCH CTL CONTRIBUTED TO THE VIOLATION. WHILE LEARNING TO FLY A NEW TYPE OF ACFT THE INSTRUCTOR SHOULD HAVE THE TASK OF TALKING TO APCH AND TWR AND REDUCE THE WORKLOAD ON THE STUDENT.

SYNOPSIS PA34 TRAINEE, WITH INSTRUCTOR, FAILED TO ADHERE TO TWR TFC PATTERN INSTRUCTIONS AFTER MAKING A TOUCH AND GO INSTEAD OF FULL STOP LNDG, AS EXPECTED. TRAINEE WANTED INSTRUCTOR TO COMPLETE ALL COM WITH ATC. COCKPIT COORD APPARENTLY INCOMPLETE.

Preparation and Remarks

The preparations made prior to this incident involve a training flight in the local area. No problem is noted with familiarity with the airport or any problems with communications noted. Visibility was a problem, since the tower had to vector the aircraft in.

From the wording in the report, in several places it can be determined that there was no cockpit coordination of duties determined before the flight. In other words, no preflight decision had been made as to who would handle radio communications and who would fly the aircraft.

Postincident Analysis

The incident resulted in no loss of aircraft separation, and there was no true runway incursion.

According to the report, the initial contact with the tower informed the controller of the intention to make touch-and-go landings (plural). The landing clearance initially included "HOLD SHORT OF RWY 36," indicating that there was either a problem in allowing a touch-and-go or that the controller had not understood that part of the request from the PA-34.

Prior to actually landing, the clearance was changed and the hold short eliminated. The aircraft proceeded to touch-and-go, after which the controller told the PA-34 it should have asked for a touch-and-go. Preceding the initial landing, the PA-34 had entered traffic to the right but had been told to enter left traffic.

The incident would be classified as pilot deviation—failure to do as instructed, a violation of the FARs. Remember, all pilot deviations result from a violation of FARs. Since no other traffic is mentioned and the hold-short instruction was canceled, there is no runway incursion severity category.

Several problems are noted in this incident, the most important of which is the lack of cockpit coordination between the instructor and the student:

Problem 1. The workload on the student learning a new airplane should not have included communications. Relieving the pilot of communications duties would have enhanced the transition training for the pilot, allowing full time and attention to flying the aircraft.

Problem 2. The assignment of duties (flying and communications) was not divided properly. In this instance, the instructor seems to be along for the ride and was doing nothing. Splitting the duties probably would have resulted in no failures to comprehend ATC instructions.

Problem 3. The controller missed the request for touch-and-goes.

Problem 4. The flight crew missed that the controller did not say anything about clearance for touch-and-go landings.

Problem 5. The student pilot became frazzled while handling an unfamiliar aircraft and concerned with reduced visibility. This confusion first showed when the traffic pattern was entered incorrectly from the right.

Lessons Learned

There are several lessons about work load and stress to be learned from this incident:

Lesson 1. It would be prudent to allow a pilot transitioning into a different aircraft to fly the airplane and not also handle radio communications. The latter can come along later as the pilot becomes more familiar and experienced with the airplane.

Lesson 2. The division of duties should be made before the flight. As in the case of two-pilot flight crews, the pilot not flying the aircraft generally handles radio communications.

Lesson 3. The flight instructor is being paid; therefore, the pilot in this case should have told the instructor to handle communications.

Lesson 4. The flight crew should have both either agreed on the instruction about entering the traffic or asked the controller for a repeat. There was a question; therefore, the controller should have been asked for a "say again."

Lesson 5. The controller should listen more clearly to what the pilot requests.

Lesson 6. The initial landing clearance included a hold-short instruction. This clearly eliminated a touch-and-go. When the clearance was amended, making the full runway length available, the pilot should have again requested a touch-and-go.

CASE 10

The Controller Didn't Listen

ASRS accession number: 441963

Month and year: June 1999

Local time of day: 1801 to 2400

Facility: ORD, Chicago-O'Hare International Airport (FIG. 4-10)

Location: Chicago, IL

Flight conditions: VMC

Aircraft 1: B727

Aircraft 2: Unknown

Pilot of aircraft 1: Captain, 11,000 hours; first officer; second officer

Pilot of aircraft 2: Unknown

Reported by: Pilot of aircraft 1

Incident description: LAHSO rejection

Incident consequence: Near collision

NARRATIVE WHEN WE WERE SWITCHED TO TWR, THEY SAID, 'CLRED TO LAND RWY 14R, HOLD SHORT RWY 27L.' I REPLIED 'CLRED TO LAND RWY 14R, UNABLE HOLD SHORT.' THE CTLR THEN SAID 'CLRED TO LAND RWY 14R.' THE CTLR THEN CLRED AN ACFT TO TAKE OFF ON RWY 27L. I STOPPED THE ACFT PRIOR TO THE INTXN, WHILE THE OTHER ACFT WAS ABOUT 50 FT OVER THE RWY. I SAID 'I TOLD YOU UNABLE HOLD SHORT.' THE CTLR SAID 'ROGER.' EITHER HE DIDN'T HEAR ME SAY 'UNABLE' OR HE IGNORED IT. THIS COULD HAVE HAD SERIOUS IMPLICATIONS—PARTICULARLY IF WE BLEW THROUGH, BALKED THE LNDG, OR IF HE REJECTED HIS TKO.

SYNOPSIS B727 FLC REJECTS LAHSO AT ORD BUT CTLR IGNORES THE UNABLE.

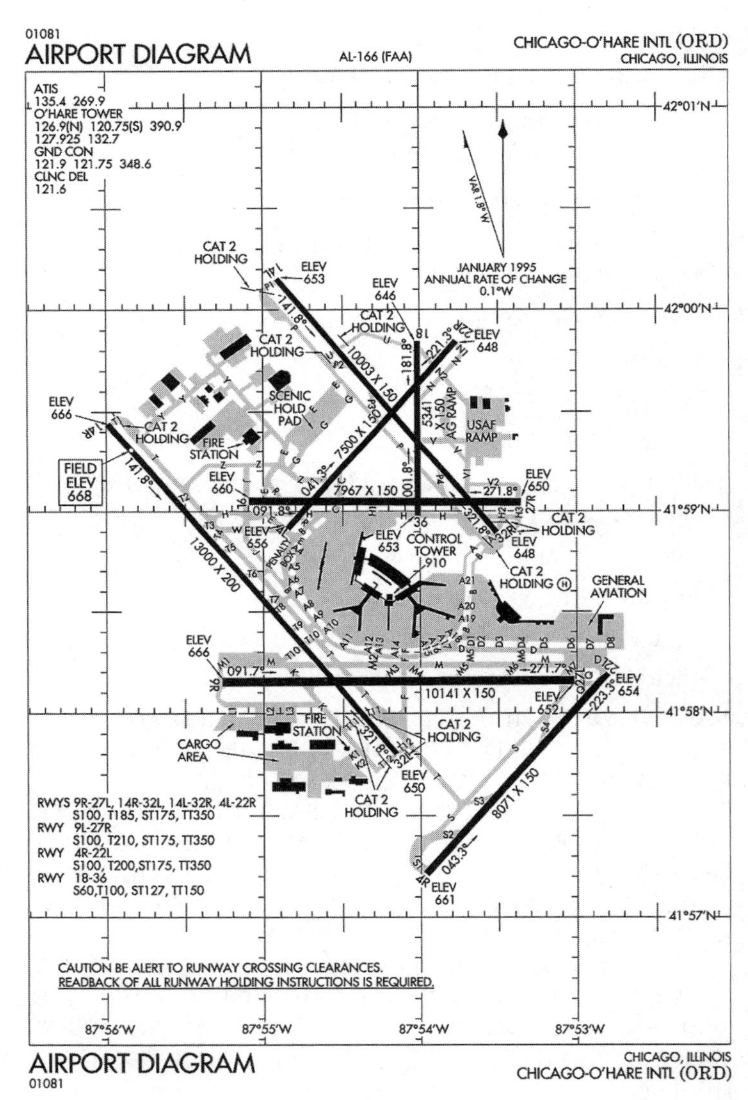

4-10 *Chicago O'Hare International, ORD (this airport diagram is not suitable for navigational purposes).*

Preparation and Remarks

The flight crew of the B727 is assumed to be completely prepared for operations at this airport. There was no indication of radio communications difficulties.

The controller does not seem to up to speed on what is happening with air traffic being controlled—an aircraft was cleared for takeoff on an intersecting runway, while another aircraft was cleared for landing on that intersecting runway.

Postincident Analysis

This incident, like many LAHSO incidents, could have been much more serious than was reported.

When the initial LAHSO clearance was rejected by the B727, the tower said, "CLRED TO LAND RWY 14R." No further instruction for hold was made. This was clearly an unlimited clearance to land. The controller then cleared an aircraft for takeoff on runway 27L—which intersects with runway 14R.

This would have been a very good time for the B727 pilot to ask ATC if they had copied the rejection for the LAHSO (regardless of the fact that the B727 had been cleared to land). This would have underlined the problem to the controller and perhaps have alerted the aircraft taking off on runway 27L of a impending problem.

As an incursion, this would have been classified as a pilot deviation, regardless of whether it was caused by operational error or not. According to the FAA's Interpretive Rule, all errors are the responsibility of the pilot. The runway incursion severity category would have been category B because the B727 did stop prior to the intersection and the other aircraft was 50 feet in the air. Had circumstances been slightly different for the landing or takeoff aircraft, the category could well have been accident!

A number of troubling problems surface from this report:

Problem 1. The flight crew rejected the LAHSO as "UNABLE HOLD SHORT," after which the controller, replied "CLRED TO LAND RWY 14R." This indicated that the controller heard and understood the rejection.

Problem 2. The controller then cleared a takeoff on runway 27L (an intersecting runway).

Problem 3. The controller did not issue a go-around to the B727 cleared for runway 14R, yet there was the potential for severe conflict if anything did not go as planned.

Problem 4. The incident was kept at ASRS report level by an observant flight crew that was able to stop short of runway 27L, allowing the other traffic through (over) at the intersection.

Problem 5. There is no mention of the other flight crew in this report, leaving open the questions of whether it was monitoring all ATC communications and visually observing airport activity.

Problem 6. The controller appears to be more concerned with moving traffic than with safety.

Lessons Learned

This incident shows how visual observation of surrounding traffic and the monitoring of airport activity are so very important.

Lesson 1. The B727 flight crew assumed that the controller understood the rejected LAHSO—the aircraft did receive a landing clearance. However, when the controller cleared for takeoff an aircraft on

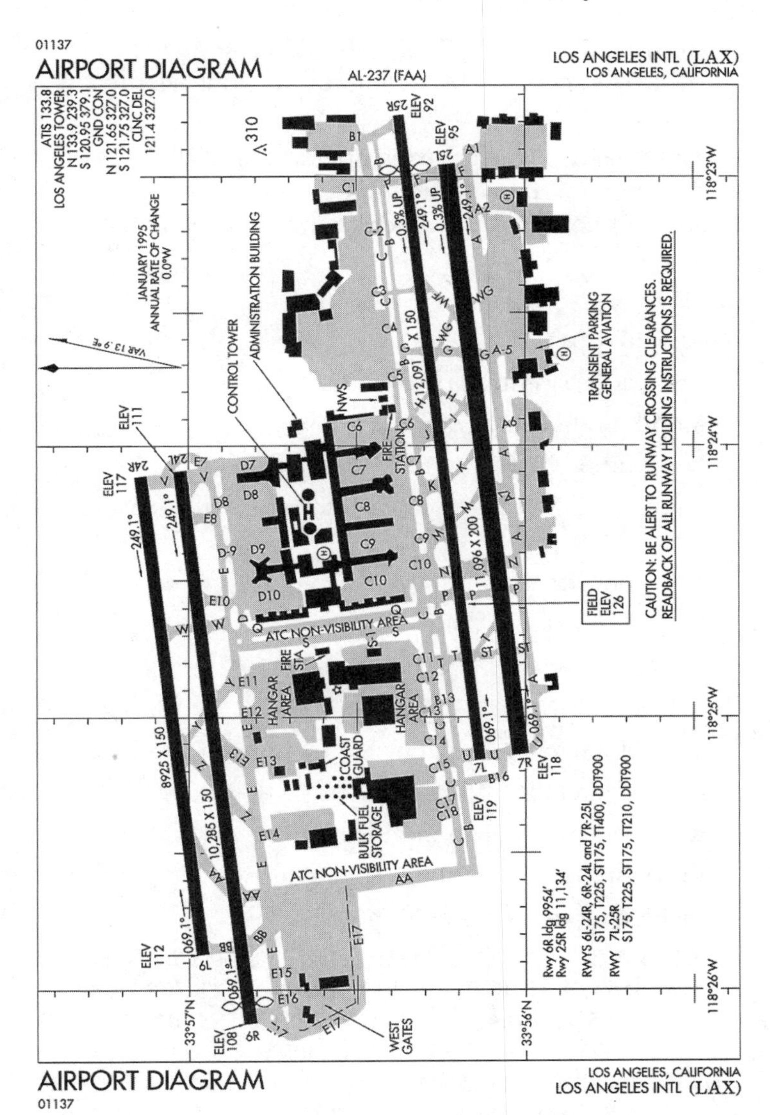

4-11 *Los Angeles International, LAX (this airport diagram is not suitable for navigational purposes).*

runway 27L, the B727 pilot should have queried the controller, perhaps by asking for the controller to "say again" the landing clearance.

Lesson 2. As always, heads up and out of the cockpit, and monitor all radio traffic. Be observant of your surroundings and the other air traffic active at the airport.

CASE 11

Who Said What

ASRS accession number: 455952

Month and year: November 1999

Local time of day: 0601 to 1200

Facility: LAX, Los Angeles International Airport (FIG. 4-11)

Location: Los Angeles, CA

Flight conditions: VMC

Aircraft 1: MD-82

Aircraft 2: B757

Pilot of aircraft 1: Captain, 15,064 hours; first officer

Pilot of aircraft 2: Unknown

Reported by: Captain of aircraft 1

Incident description: Runway incursion

Incident consequence: Near collision

NARRATIVE I WOULD LIKE TO INFORM YOU OF THE FACTS THAT OCCURRED AT LAX ARPT. AFTER WE LANDED ON RWY 25L AT LAX, THE TWR CTLR INSTRUCTED US DURING THE LNDG TO CLR THE RWY 25L VIA TXWY N AND CROSS RWY 25R AND REMAIN ON TWR FREQ. THESE INSTRUCTIONS WERE READ BACK BY THE FO WITHOUT RECEIVING FURTHER COMS FROM THE TWR CTLR. WHILE BEING AT TXWY N AND AFTER THE RWY 25R WAS CLRED, WE STARTED XING ACCORDING TO THE INSTRUCTIONS

AND AT THE MIDDLE OF THE RWY 25R, THE TWR CTLR THEN GAVE US THE INSTRUCTION TO HOLD SHORT OF RWY 25R. WE COULDN'T COMPLY WITH THE INSTRUCTION BECAUSE THE AIRPLANE WAS IN THE MIDDLE OF THE RWY, AS WE MENTIONED BEFORE. THAT'S WHY WE EXPEDITED XING THE RWY. AT THAT MOMENT WE SAW ABOVE US AN ACFT, APPARENTLY A B757. LATER THE TWR CTLR SAID THAT THE INSTRUCTION WAS TO HOLD SHORT OF THE RWY AND REMAIN ON FREQ. THE FO TOLD THEM THAT HE READ BACK THE INSTRUCTIONS TO CLR THE RWY 25L VIA TXWY N AND CROSS RWY 25R REMAINING ON TWR CTL FREQ. AFTER THAT, THE TWR CTLR INSTRUCTED US TO CHANGE THE FREQ AND THEY GAVE US A PHONE NUMBER TO CONTACT THEM.

SYNOPSIS DIFFERENCE OF OPINION AS TO CLRNC TO CROSS RWY AT LAX.

Preparation and Remarks

The flight crew was prepared for operations at LAX, and there are no indications of communications difficulties. They received and read back their clearance as required. There were no indications of concerns about position or any other indications of position confusion.

Postincident Analysis

From the report it is understood that the aircraft was cleared to leave runway 25L at taxiway N and to cross runway 25R. The clearance was read back to the controller. The controller did not acknowledge the readback, nor did the MD-82 challenge the controller for not acknowledging.

When the controller gave hold-short instructions to the aircraft, the aircraft already was on the runway. At that time, a B757 passed over them. In reading the report, the hold-short instruction seems more like a last-minute amendment to a previously issued clearance.

The incursion is classified as pilot deviation, and the runway incursion severity category would be category A, based on the statement, "AT THAT MOMENT WE SAW ABOVE US AN ACFT, APPARENTLY A B757." Had the taking off aircraft not been in the air, they would have been in the middle of the MD-82.

This could have been a very serious incident, rather than an ASRS report. Several problems are noted:

' *Problem 1.* The controller failed to acknowledge the read-back, at which time the flight crew should have said, "PLEASE CONFIRM MY CLEARANCE AS...," thereby placing the burden on the controller (as much as possible, keeping in mind the Interpretive Rule).

Problem 2. There is no indication that the MD-82 flight crew was monitoring other radio communications at the airport or it would have heard the takeoff clearance for the B757.

Problem 3. There is no indication that members of the flight crew had their heads out of the cockpit, save for the statement about the B757 passing over them. They might have been able to see the B757 as they (MD-82) approached runway 25R on taxiway N.

Problem 4. Short of playing back all the tapes [ATC and the cockpit voice recorder (CVR)], it is not possible to determine who said or didn't say exactly what. Therefore, the difference of opinion as to the clearance will stand. Such playbacks generally are used only when the National Transportation Safety Board (NTSB) is cleaning up the aftermath of a crash or collision—when it is too late.

Lessons Learned

The lessons learned from this incident are intended to reduce laxness on the part of flight crews:

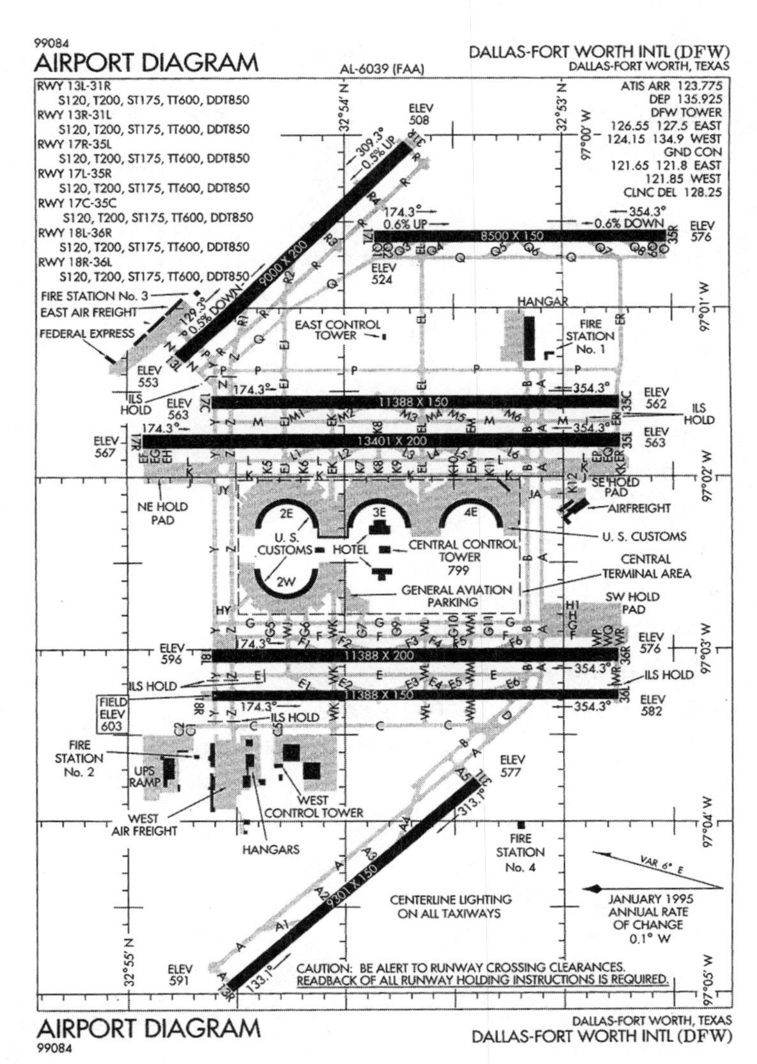

4-12 *Dallas-Fort Worth International, DFW (this airport diagram is not suitable for navigational purposes).*

Lesson 1. When a clearance is read back to the controller, expect and get an acknowledgment. If you get no acknowledgment, you do not know if it was ever received. A recommended procedure is to say, "Please confirm my clearance as..."

Lesson 2. Again, keep your head out of the cockpit, and watch for airport activity. Be particularly vigilant when crossing runways.

Lesson 3. Again, monitor other radio traffic at the airport, listening for clearances involving runways you will be using and/or crossing.

Both Lessons 2 and 3 are very basic, yet they have saved countless numbers of lives. Be vigilant, and be observant.

CASE 12

No LAHSO Read-Back

ASRS accession number: 413556

Month and year: September 1998

Local time of day: 0601 to 1200

Facility: DFW, Dallas–Fort Worth International Airport (FIG. 4-12)

Location: Dallas–Fort Worth, TX

Flight conditions: VMC

Aircraft 1: B757

Pilot of aircraft 1: Captain

Reported by: Controller

Incident description: Failure to read back hold-short instruction

Incident consequence: ASRS report made.

NARRATIVE ACR XX B757 WAS CLRED TO LAND RWY 35C WITH RESTR TO HOLD SHORT OF TXWY EJ ON LNDG ROLL. HE DID NOT READ BACK THE HOLD SHORT

INSTRUCTIONS. THIS IS HAPPENING ON A REGULAR BASIS WITH ACR PLTS. I BELIEVE IT'S THEIR WAY OF PROTESTING THE FACT THAT THEY DON'T LIKE LAHSO PROCS. IF THEY DON'T WANT TO ACCEPT THE CLRNC, THEN THEY NEED TO SAY SO—NOT JUST IGNORE IT. THIS BEHAVIOR GREATLY INCREASES THE POTENTIAL FOR SYS ERRORS.

SYNOPSIS A DFW CTLR GIVES A LAHSO INSTRUCTION TO A LNDG B757 BUT RECEIVES NO ACKNOWLEDGMENT OF THE CLRNC.

Preparation and Remarks

The controller was prepared and issued a LAHSO clearance. The flight crew failed to read back the clearance as required. This type of problem was relatively prevalent for a period of time when pilots were protesting LAHSO clearances. Currently, it has abated because it has become more acceptable to just refuse the LAHSO clearance—which is the aircraft commander's prerogative.

Postincident Analysis

For reasons of workload and increased air traffic at busier airports, LAHSO clearances have become quite common. In this incident, the flight crew failed to read back the LAHSO clearance—a read-back is required. There is, however, no indication that the controller challenged the B757 for a read-back.

This report does not reflect an actual runway incursion but rather a dangerous attitude among some pilots toward the acceptance of LAHSO clearances. It is a pilot deviation because the pilot violated the FARs by not making a read-back of the LAHSO clearance.

Several problems surround this incident:

Problem 1. LAHSO clearances are not popular with pilots. However, the pilots do have the option of refusing a LAHSO clearance. No controller can force a pilot to accept a LAHSO clearance. However, the pilot must state that the LAHSO clearance is being rejected—not just ignore it.

Problem 2. The controller did not challenge the B757 about the missing read-back. The controller should have said, "Aircraft xxx acknowledge and read back your landing clearance." This would have left no doubt in the pilot's mind that a read-back was necessary.

Problem 3. The controller's statement about increasing the potential for system errors is right on target. This type of incident only makes for additional confusion and danger. If it were, in fact, the pilot's intention to just ignore the LAHSO clearance, then the behavior is childish and unacceptable—far less than the standards that pilots generally are measured by.

Lessons Learned

The underlying lesson to be learned here is the aircraft commander's responsibility (requirement) to ensure that the LAHSO clearance is read back. If the clearance is not read back, then the responsibility is to inform the controller that the clearance is rejected. The requirement of read-back leaves no room for misunderstandings. The clearance was either accepted and read back or rejected.

Failure to read back is a serious issue and cannot be allowed to go uncorrected. If the pilot fails to make the read-back, the controller must demand that a read-back be made.

CASE 13

The Tired Pilot

ASRS accession number: 455679

Month and year: November 1999

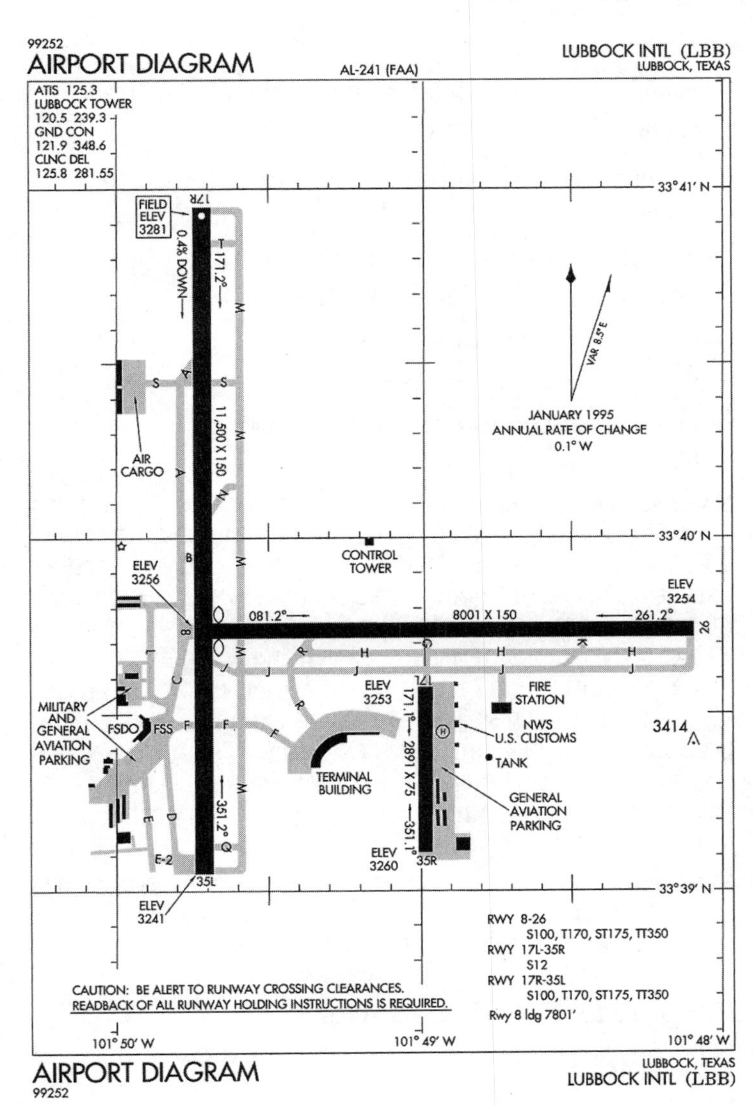

4-13 *Lubbock International, LUB (this airport diagram is not suitable for navigational purposes).*

Local time of day: 1801 to 2400

Facility: LBB, Lubbock International (FIG. 4-13)

Location: Lubbock, TX

Flight conditions: VMC

Aircraft 1: Cessna 172 Cutlass

Aircraft 2: Unknown jet transport

Pilot of aircraft 1: Single pilot, 136 hours

Pilot of aircraft 2: Unknown

Reported by: Pilot of aircraft 1

Incident description: Runway incursion

Incident consequence: None

NARRATIVE AFTER LNDG ON RWY 26, I WAS TOLD TO TAXI TO END OF RWY 26, HOLD SHORT OF RWY 17. I WAS WATCHING FOR MARKINGS ON RWY TO INDICATE WHERE I SHOULD STOP. STOPPED AT END OF RWY, PLACING ME AT EDGE OF RWY 17. OTHER ACFT (LARGE JET) WAS DOING TOUCH-AND-GOES. IF I HAD GONE EVEN A FEW FT FURTHER, THERE COULD HAVE BEEN A COLLISION. NEED TO STOP MUCH EARLIER, AND BE MORE FAMILIAR WITH TAXIING ON RWYS. CONTRIBUTING FACTORS: 1) SEVERAL TXWYS UNDER REPAIR, MAKING TAXI ON RWY MORE LIKELY 2) LACK OF EXPERIENCE TAXIING ON RWY 3) FATIGUE AFTER XCOUNTRY FLT AT NIGHT.

SYNOPSIS PVT PLT OF A C172 STOPPED PAST THE HOLD SHORT AFTER LNDG TO A CONVERGING RWY CAUSING A POTENTIAL CONFLICT WITH A JET ACFT LNDG.

Preparation and Remarks

Lubbock is not a very busy airport. The aircraft involved were a Cessna 172 and a transport-type jet (doing touch-and-go landings). The Cessna pilot was not familiar with the airport, as stated in the report, "AND BE MORE FAMILIAR WITH TAXIING ON RWYS," and probably was not using an airport diagram. Radio communications for the airport were not mentioned and are assumed to have been working properly.

Postincident Analysis

The incident was caused by the pilot of the Cessna not stopping and holding where ATC had instructed: "TOLD TO TAXI TO END OF RWY 26, HOLD SHORT OF RWY 17." The pilot reports looking for, but not seeing, runway surface marking that indicated where to stop. Due to construction, taxiing was done on the runway, and it is possible that proper markings were not present. However, had the Cessna pilot been monitoring the airport's other radio traffic, the clearance for the transport would have been noted and the aircraft watched for. In addition, the pilot indicated being fatigued after a nighttime cross-country flight.

Taxiing on the runway, rather than on a proper taxiway (closed due to construction), should not have been a surprise for the pilot, since a check of the NOTAMs would have disclosed this information ahead of time.

The controller should have been visually monitoring the Cessna as it rolled out and should have instructed the pilot with a fast reminder to stop when it became obvious the plane was taxiing too close to runway 17.

The incident would be classified as a pilot deviation, and the runway incursion severity category would be category B because there was decreased separation and a significant potential for a collision (had the Cessna not stopped at the edge of the runway).

The problems noted from this report indicate a lack of attention to details and surrounding airport activities:

Problem 1. The pilot saw no lines to use as a hold point at the end of runway 26.

Problem 2. Had the pilot checked the NOTAMs, the taxiway closures would have been known in advance.

Problem 3. The controller does not appear to have noticed this incursion to the active runway.

Lessons Learned

Flying requires full time and attention—there is no room for anything less. A great attention to details must be made at all times to prevent incidents such as this from happening in the first place.

The lessons learned from this incident all deal with paying attention to details.

Lesson 1. Know the layout of the airport. If you are unfamiliar with the airport, use an airport diagram. The appearance of an intersecting runway is visually obvious to the left and to the right, and the use of an airport diagram would have given clues as to where on runway 26 the Cessna was as it progressed toward runway 17. There would have been signs indicating the various taxiways intersecting to the left and to the right.

Lesson 2. Always check the NOTAMs for your destinations and for likely stopping points along the route.

Lesson 3. Listen to the radio traffic. In this case, the Cessna pilot should have been aware of the transport landing on runway 17, seen the aircraft coming in on final, and stopped well in advance of the runway intersection without the need for signs.

Lesson 4. Be rested when flying. There is no place for fatigue in the cockpit of an airplane—or cab of the tower for that matter.

Lesson 5. Controllers must be aware of all aircraft activity on the airport that may affect or be affected by their actions.

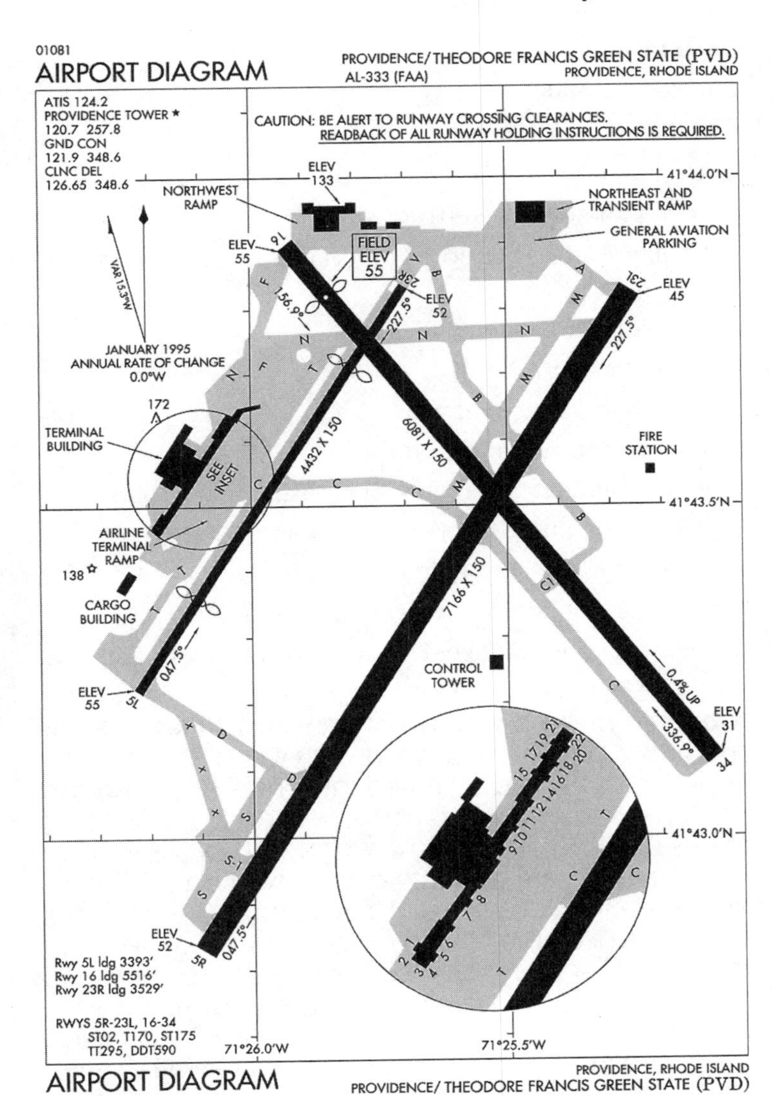

4-14 *Providence/Theodore Francis Green State Airport, PVD (this airport diagram is not suitable for navigational purposes).*

CASE 14

The Lost Captain

ASRS accession number: 456996

Month and year: December 1999

Local time of day: 1801 to 2400

Facility: PVD, Theodore Francis Green State Airport (FIG. 4-14)

Location: Providence, RI

Flight conditions: IMC

Aircraft 1: Large transport

Aircraft 2: Large transport

Pilot of aircraft 1: Captain; first officer, 4300 hours

Pilot of aircraft 2: Captain; first officer, 6200 hours

Reported by: Pilot of aircraft 1

Incident description: Runway incursion

Incident consequence: FAA reviewed incident with flight crew.

NARRATIVE WE LANDED ON RWY 5R IN PROVIDENCE (PVD). WE DID A CAT II APCH AUTOLAND RVR TOUCHDOWN 1200 FT ROLLOUT 1400 FT. WHILE STILL ON THE RWY THE CTLR GAVE US OUR TAXI CLRNC: L TURN N CROSS RWY 16, LET ME KNOW WHEN YOU CROSS RWY 16, STAY WITH ME. WE TAXIED OFF THE RWY ONTO TXWY N. WHILE TAXIING I THEN TOLD THE CAPT I WAS OFF OF #1 RADIO AND GOING OVER TO COMPANY TO GET THE GATE INFO. I CALLED COMPANY AND HAD TO WAIT FOR A MIN FOR THEM TO ANSWER THEY CAME BACK AND TOLD ME THE GATE. I THEN TOLD THE CAPT I WAS BACK ON RADIO #1. AT THAT TIME I WAS NOT SURE WHERE WE WERE AND WAS LOOKING FOR A TAXI SIGN. I THEN ASKED THE CAPT AND HE DIDN'T ANSWER. I THEN SAW THAT THE BLUE TAXI EDGE LIGHTS WERE DISAPPEARING AND WE WERE ENTERING WHITE LIGHTS. I THEN TOLD

THE CAPT TO STOP THE PLANE. I BELIEVED WE WERE ENTERING A RWY. HE THEN WENT TO TURN R TO GET OFF, BUT I TOLD HIM TO STOP BECAUSE THERE WAS GRASS ON THE R SIDE OF ME AND WE WOULD GO INTO THE GRASS. AT THAT POINT THE CAPT GOT ON THE RADIO AND TOLD THE TWR WE WERE ON RWY 23R. I THEN TOLD THE CAPT THAT I COULD SEE A SIGN RWY 23L RIGHT IN FRONT OF ME WITH A TXWY K SIGN OFF MY R. THE CTLR THEN CLRED AN AIR JET TO TAKE OFF OF RWY 5R. I CAME ON THE RADIO AND TOLD THE CTLR WE WERE ON RWYS 5R/23L. THE CAPT CAME ON THE RADIO AND TOLD THE CTLR WE WERE AT RWY 23R. (I WAS TRYING TO SHOW THE CAPT THE SIGN THAT SAID RWY 23L AND TO EXPLAIN TO HIM WHERE WE WERE AT.) THE CTLR THEN CLRED THE ACR A JET TO TAKE OFF RWY 5R AGAIN. THE ACR A ASKED THE CTLR WHERE THE ACR B JET WAS AND THE CTLR TOLD THEM WE WERE AT RWY 23R AND THAT WAS NOT AN ACTIVE RWY IN THIS KIND OF WX. I CAME BACK ON THE RADIO AND TOLD THE CTLR WE WERE AT RWYS 5R/23L. AT THAT POINT THE ACR A JET SAID THEY WERE NOT TAKING OFF UNTIL 'WE KNEW WHERE ACR B IS.' I THEN TOLD THE CTLR I HAD A SIGN FOR RWY 23L AND A TAXI SIGN ON MY R AND L. SHE THEN TOLD US TO TAXI TO RWY 16L TXWY C TO THE GATE AND LET ME KNOW WHEN YOU ARE AT THE GATE. WE TAXIED AND I TOLD THE TWR WHEN WE WERE AT THE GATE. I BELIEVE THE PROB WAS CAUSED BY LOW VISIBILITY, RVR 1200 FT POOR LAYOUT OF ARPT SIGNS NOT WELL LIT, THE FACT THAT I CALLED THE COMPANY AND PUT MY HEAD DOWN WHEN I SHOULD HAVE STAYED WITH THE CAPT TAXIING, ALSO WHEN THE CAPT WASN'T SURE HE SHOULD HAVE STOPPED AND ASKED ME OR THE TWR AND NOT CONTINUED TO TAXI. I SHOULD NOT HAVE PUT MY HEAD DOWN TO CALL THE COMPANY ESPECIALLY IN THE LOW VISIBILITY. I SHOULD HAVE STAYED WITH THE CAPT WHILE HE WAS TAXIING. HAD I DONE THAT I WOULD HAVE SEEN HE WAS TAKING A WRONG TURN. ALSO THE CAPT SHOULD HAVE STOPPED

AND ASKED WHEN HE WAS NO LONGER SURE WHERE HE WAS. TO PREVENT A RECURRENCE, 1) BETTER ARPT SIGNS LIGHTING, 2) BOTH OF YOU STAY HEADS OUT AND TAXIING TILL AT THE APRON, 3) WHEN NOT SURE OF TAXIING INSTRUCTION OR WHERE YOU ARE AT, STOP THE PLANE AND ASK FOR INSTRUCTION.

SYNOPSIS RWY INCURSION AT PVD DURING RESTR VISIBILITY CONDITIONS.

Preparation and Remarks

On the date and time of this incident, PVD was operating under instrument conditions, and visibility was very poor. The three parties, two aircraft and the controller, are assumed to have been prepared with the necessary materials and knowledge to operate at this airport. Radio communications and equipment were operating normally.

Postincident Analysis

The incident was caused by the pilot of the transport aircraft becoming lost on the airport. Contributing factors to this incident were very poor visibility and the first officer not being heads up with the captain. Another factor is the captain's attitude of arguing about the aircraft's location.

Confusion about the location on the airport diagram may have been caused when the aircraft mistakenly turned left onto taxiway M rather than continuing on taxiway N as instructed. The intersection of taxiways M and K with runway 16/36 may have appeared similar to taxiway N and runways 16/36 and 23R/5L.

The controller should not have cleared the second transport to take off until it was known if the runway was clear. The flight crew of the second aircraft acted properly in refusing to move until the first aircraft's location had been determined.

This incident would be classified as a pilot deviation, and the runway incursion severity category would be category C because there was ample time and distance to avoid a potential collision.

The problems noted from this incident involve a lack of attention to the outside during ground movement, a lack of highly visible airport signage, and a lack of controller attention to the situation:

Problem 1. The first officer should have been heads up. By the time he returned from talking on the company radio, the report says, "AT THAT TIME I WAS NOT SURE WHERE WE WERE AND WAS LOOKING FOR A TAXI SIGN." He was no help to the already lost captain at this point.

Problem 2. "SIGNS NOT WELL LIT" indicates that taxiway signs were not easily identified due to poor visibility conditions. They should be lighted better for improved visibility.

Problem 3. The controller should not have cleared the second transport for takeoff on any runway until the true whereabouts of the first aircraft was known.

Lessons Learned

Taxiing in poor visibility conditions requires full attention to the outside of the aircraft. It also requires that airports signs and surface markings be of such a design as to be fully visible during periods of poor visibility.

Lesson 1. The first officer states this very well by saying, "BOTH OF YOU STAY HEADS OUT AND TAXIING TILL AT THE APRON, 3) WHEN NOT SURE OF TAXIING INSTRUCTION OR WHERE YOU ARE AT, STOP THE PLANE AND ASK FOR INSTRUCTION."

Lesson 2. There is a need for highly visible signage.

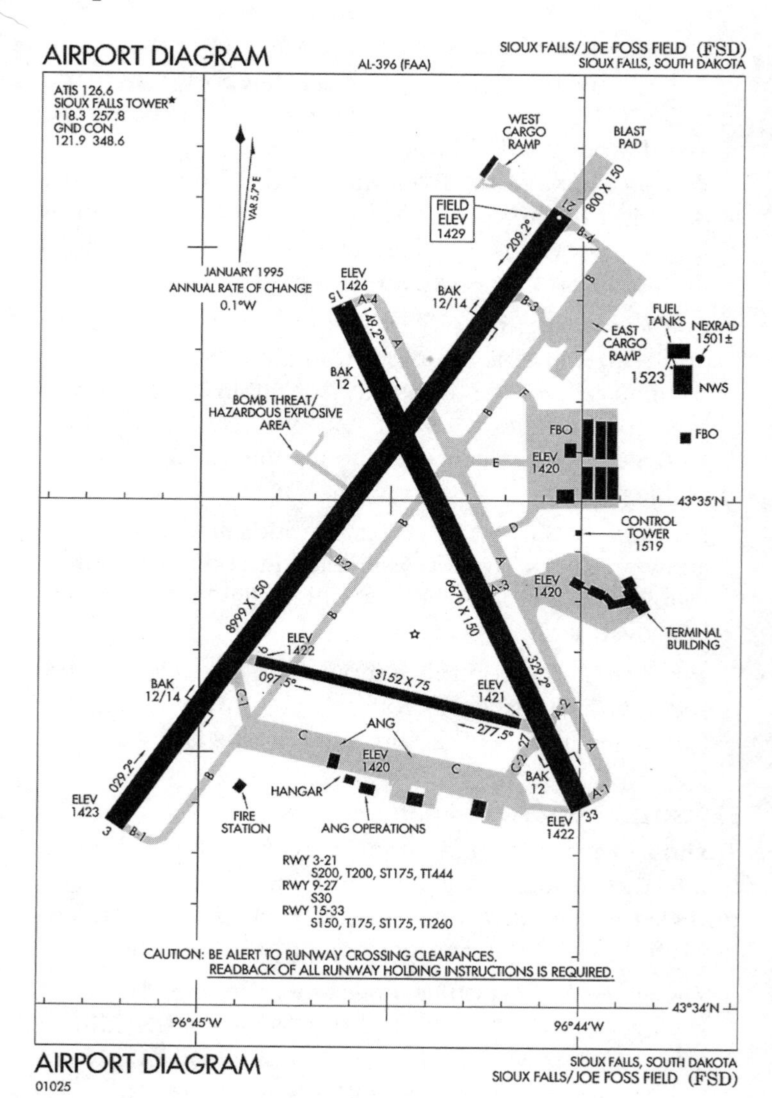

4-15 *Sioux Falls Field, FSD (this airport diagram is not suitable for navigational purposes).*

Lesson 3. Always monitor the radio for other airport activity. This is exactly what the second aircraft did and the reason for their not starting their takeoff roll until the first aircraft had been positively located.

Lesson 4. No aircraft should be cleared for any airport movement until the lost aircraft has been located.

CASE 15

The Mowing Machine

ASRS accession number: 409053

Month and year: July 1998

Local time of day: 1201 to 1800

Facility: FSD, Joe Foss Field (FIG. 4-15)

Location: Sioux Falls, SD

Flight conditions: VMC

Aircraft 1: Cessna 152

Pilot of aircraft 1: Single pilot, 150 hours

Reported by: Pilot of aircraft 1

Incident description: Runway incursion

Incident consequence: Pilot took evasive action.

NARRATIVE ON SHORT FINAL AT FSD, I WAS CLRED TO LAND RWY 21 AND HOLD SHORT OF RWY 33. JUST PRIOR TO LNDG, A MOWER PULLED OUT RIGHT AT THE END OF RWY 21. AFTER ADDING PWR TO ENSURE CLRING THE MOWER, I WAS UNABLE TO HOLD SHORT OF RWY 33 AND EXECUTED A TOUCH AND GO. RECOMMEND THAT CTLRS DO NOT ADVISE PLTS TO HOLD SHORT WHEN EQUIP IS ON OR AROUND THE RWY.

SYNOPSIS SMA PLT EXECUTES MANEUVER DUE TO ARPT MOWER RWY TRANSGRESS. PLT BELIEVES TWR LCL CTLR WAS SOMEHOW INVOLVED.

Preparation and Remarks

The incident involved a LAHSO, for which the pilot was prepared. The pilot of the Cessna should have been able to see the mower working around the runway while on final and should have been prepared to take evasive action. There were no problems with radio communications noted in the report.

Postincident Analysis

The incident was caused by the mowing equipment entering on the runway on which the aircraft was landing. The Cessna was able to make a touch-and-go of the landing; however, had there been an aircraft crossing the intersection of runways 21 and 33, there could have been a collision. This was an emergency situation in the eyes of the pilot, allowing the specific maneuver made to avoid the mower. It was the pilot's call per FAR Section 91.3 (b): "In an in-flight emergency requiring immediate action, the pilot in command may deviate from any rule of this part to the extent required to meet that emergency."

This incursion was caused by a vehicle/pedestrian deviation, the entry or movement on the runway or taxiway area of a vehicle or pedestrian that has not been authorized by ATC. It is assumed that ATC did not clear the mower onto the runway in front of the landing aircraft.

The problems from this incident involve the lack of observation of the mower and the lack of control of the mowing operation:

Problem 1. The Cessna pilot should have seen the mower while on final and queried ATC if uncomfortable about its location and direction of travel.

Problem 2. The controller should have been monitoring the activities of the mowing machine

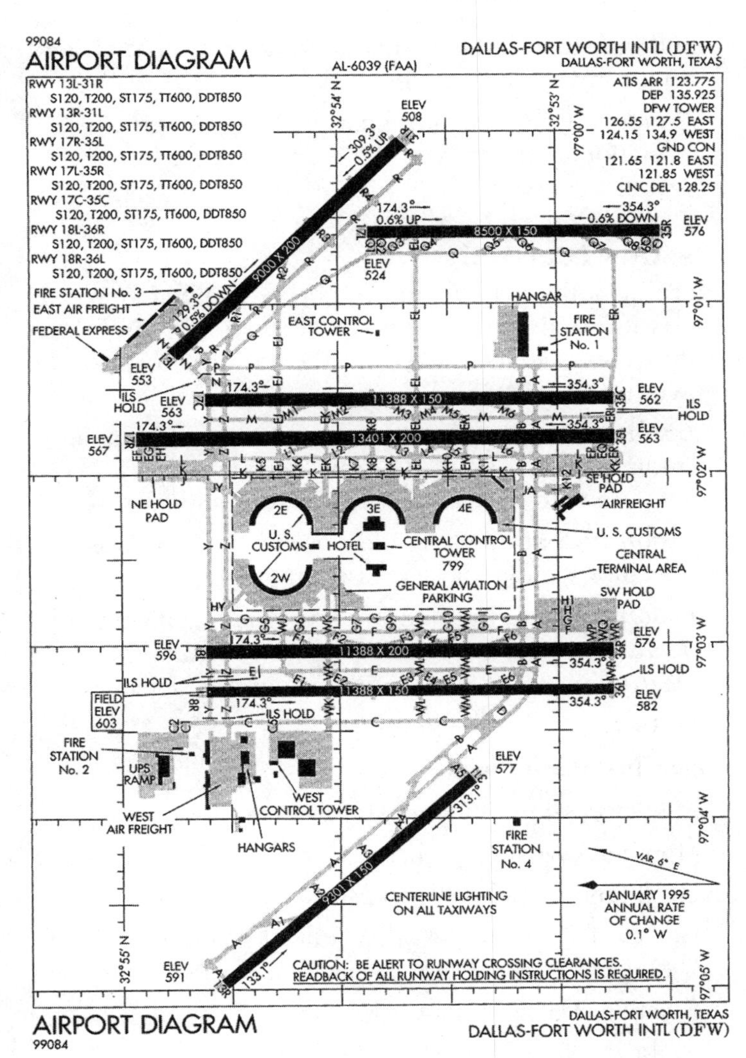

4-16 *Dallas-Fort Worth International, DFW (this airport diagram is not suitable for navigational purposes).*

and considered this when clearing an aircraft to land on a runway close to the mower.

Problem 3. Who was in charge of the mowing operation, and why was the mower allowed to turn onto the runway?

Lessons Learned

There is nothing more important in flying than visual observation. Everyone has to look, look, and look again. This means the pilot, the controller, and the mower operator.

Lesson 1. Observe the airport and runway as you are landing. Look for any abnormal activity that may interfere with the safe landing of your aircraft.

Lesson 2. Controllers should observe maintenance operations and avoid use of nearby runways.

Lesson 3. The operator of the mowing machine should be given a copy of *Airport Ground Vehicle Operations,* an FAA guide, to read and understand.

CASE 16

Roger, Just Plain Roger

ASRS accession number: 414679

Month and year: September 1998

Local time of day: 1801 to 2400

Facility: DFW, Dallas–Fort Worth International Airport (FIG. 4-16)

Location: Dallas–Fort Worth, TX

Flight conditions: VMC

Aircraft 1: B757

Pilot of aircraft 1: Captain; first officer

Reported by: Controller

Incident description: Runway incursion

Incident consequence: None

NARRATIVE ACR X, A B757, CHKED IN. THE TWR CLRED HIM TO LAND RWY 35C WITH RESTR TO HOLD SHORT OF TXWY EJ. ACR X ACKNOWLEDGED WITH 'CLRED TO LAND' (NO HOLD SHORT READ BACK). THE TWR THEN SAID 'AND HOLD SHORT OF TXWY EJ FOR XING TFC.' ACR X RESPONDED WITH 'ROGER'—AGAIN WITH NO READ BACK. PLTS HAVE MADE IT KNOWN THAT THEY DON'T LIKE LAHSO, SO THEY'RE PLAYING GAMES ON FREQ (NOT ALL PLTS). THERE ARE GOING TO BE SAFETY PROBS IF THIS CONTINUES. IF THEY DON'T WANT TO ACCEPT THE HOLD SHORT INSTRUCTIONS, THEY NEED TO ADVISE.

SYNOPSIS A B757 WAS GIVEN A LNDG CLRNC BY THE TWR AND ASKED TO HOLD SHORT OF TXWY EJ. THE PLT ACKNOWLEDGED THE LNDG CLRNC, BUT NOT THE HOLD SHORT INSTRUCTIONS. THE CTLR STATED AGAIN TO HOLD SHORT OF THE TXWY AND THE PLT RESPONDED WITH A 'ROGER.'

Preparation and Remarks

The controller was prepared for the landing of the B757, and the flight crew is assumed to have been ready as well. The problem involving this incident is the acceptance of LAHSO clearances by pilots. LAHSO clearances are controversial, and acceptance is always at the discretion of the pilot. No problems were noted in radio communications, i.e., no indication of missed transmissions or interruptive malfunctions.

Postincident Analysis

This incident is an example of an aircraft and its passengers endangered by the actions of the pilot. There is no question that the aircraft commander has the authority to accept or refuse a LAHSO clearance. However, there is a requirement that clearances be read back and for both the controller and the pilot to correct errors. Note that all errors are the responsibility of the pilot for incorrect and/or uncorrected read-backs.

This being said, this is an example of a pilot just not wanting to make things run smoothly. The statement in the report accurately reflects this: "PLTS HAVE MADE IT KNOWN THAT THEY DON'T LIKE LAHSO, SO THEY'RE PLAYING GAMES ON FREQ (NOT ALL PLTS). THERE ARE GOING TO BE SAFETY PROBS IF THIS CONTINUES."

The pilot in question had only to decline the LAHSO, and other accommodations would have been made for the landing. This incursion is pilot deviation, and the runway incursion severity category is category D because there is no indication of crossing traffic and therefore little chance of collision.

The problems noted with this incursion have to do with pilot attitude:

Problem 1. Commercial pilots are professionals. They are tasked with carrying passengers and freight safely. Controllers are professionals. They are tasked with maintaining safe aircraft separation during airport operations. Cooperation between the two is essential for each to carry out their tasks.

Problem 2. A lack of FAA enforcement actions in cases such as this is exemplified by this report. Specifically, the pilot has the right to decline the LAHSO clearance but not the right to ignore radio communications. If the FAA were to sanction a few pilots for this behavior, most likely it would stop.

Lessons Learned

There is a lesson to be learned from this report about cooperation—specifically, a little of it goes a long way.

Lesson 1. Make the controller aware of your decision to refuse a LAHSO clearance. Failure to do so not only makes the controller's job more difficult but also endangers the passengers on your aircraft and those on aircraft using intersecting runways.

As a postscript to this incident, an old friend once said, "You can't fight city hall. But if you cooperate, then city hall will fight for you." Something to think about.

Lessons Learned in this Chapter

This chapter has closely examined a number of different arrival-type runway incursions and other arrival incursion-type scenarios, as reported in the ASRS reports used for the case studies. Pilot deviations and operational errors were shown. Recognize that aircrews can do little about operational errors, except for vigilance and bringing these errors to the attention of ATC.

The FAA's Interpretive Rule was shown by example, and the fact that the rule generally places all blame for errors on the pilot was explained. For more details about the Interpretive Rule, see Appendix B.

In each case study, several lessons were learned. Of these lessons, the following appear to be the most common:

Visual vigilance. Keep your heads out of cockpits and look around. Watch for other aircraft activity on the airport. On final, watch for taxiing aircraft (or equipment) that may enter the active runway. When taxiing across runways, watch for aircraft landing and taking off (their lights are on to make them easier to see).

Radio vigilance. Listen to all of ATC's radio traffic, not just radio traffic directed to you alone. This provides the information about other aircraft landing and taking off at the airport.

Airport diagrams. Airport diagrams will aid you in visualizing your approach to the correct runway, to find your assigned taxiway, and to see exactly where you may be required to hold short. Reference the diagrams well prior to final approach, and remember to keep your head out of the cockpit.

Read back clearances. All clearances containing hold instructions and all LAHSO clearances must be read back. Read-back is an excellent means of avoiding both pilot deviations and operational errors.

At anytime you have a problem, either with the landing clearance as received or while taxiing after landing, contact ATC for help. If you do not understand a clearance, say either "say again" or "words twice" to get a repeat. If you have received a LAHSO clearance, the acceptance is completely at the discretion of the aircraft commander.

5

Uncontrolled Airport Runway Incursions

As stated in the *Aeronautical Information Manual* (AIM), "There is no substitute for alertness while in the vicinity of an airport. It is essential that pilots be alert and look for other traffic and exchange traffic information when approaching or departing an airport without an operating control tower. This is of particular importance since other aircraft may not have communication capability or, in some cases, pilots may not communicate their presence or intentions when operating into or out of such airports. To achieve the greatest degree of safety, it is essential that all radio-equipped aircraft transmit/receive on a common frequency identified for the purpose of airport advisories."

Uncontrolled airports run the gamut from grass strips to multirunway operations with all services, including a flight service station (FSS), but no control tower or a tower that only operates during specific hours. In general, uncontrolled airports are marked, signed, and lighted, as are controlled airports. The only real difference is that you have no one responsible for issuing clearances for taxi and runway usage. Radio communications is through the FSS or UNICOM or by making self-announcements of your location and intentions so that others may hear you.

Radio communications (or announcements) at an airport without an operating control tower are done by use of a common traffic advisory frequency (CTAF). The frequency may be that of a UNICOM, MULTICOM, FSS, or the tower frequency when it is closed. The CTAFs are listed in the *Airport/Facilities Directory*.

Pilots of inbound aircraft should monitor and communicate as appropriate on the designated CTAF from 10 miles out through entering the traffic pattern and final approach and report leaving the runway. Pilots of departing aircraft should monitor and communicate from startup, during taxi, before taking the active runway for takeoff, and until 10 miles away from the airport.

Placing the Blame

When studying runway incursions that happened at controlled airports, one of the classifications of error was operational error. Since tower controllers are not involved at uncontrolled airports, this leaves only pilot deviations and vehicle/pedestrian deviations as possible classifications. The use of runway incursion categories will be applied to the following case studies as appropriate.

For more complete details about flying at uncontrolled airports, read Advisory Circular AC No. 90-42F, found in Appendix C of this book.

CASE 1

Two Wrongs Make a Right

ASRS accession number: 438326

Month and year: May 1999

Local time of day: 1201 to 1800

Facility: UDD, Bermuda Dunes Airport (FIG. 5-1)

Location: Palm Springs, CA

Flight conditions: VMC

Aircraft 1: GA type single

Aircraft 2:GA type twin

Pilot of aircraft 1: Single pilot, 4500 hours

Pilot of aircraft 2: Unknown

Reported by: Pilot of aircraft 1

Incident description: Runway incursion

Incident consequence: Evasive action

NARRATIVE WHAT REALLY CAUSED THE PROB WAS LACK OF COM. RWY AT BERMUDA IN USE WAS RWY 10. LIGHT TFC WAS USING RWY 10, HOWEVER, THE WIND WAS SUCH THAT EITHER RWY 10 OR RWY 28 WOULD WORK. I TAXIED OUT CALLING FOR DEP ON RWY 10. NEXT XMISSION WAS GOING INTO POS ON RWY 10 DEPARTING DOWNWIND. PRIOR TO MY MOVING INTO POS, I HEARD 2 CALLS—MAYBE 3, BUT IT WAS CARRIER SIGNAL ONLY. I ANNOUNCED SAYING "CARRIER ONLY." I ROLLED ON TKO, HEARING NO FURTHER XMISSIONS. NEXT I SEE A TWIN

5-1 *Bermuda Dunes Airport (this airport diagram is not suitable for navigational purposes).*

ROLLING AT ME FROM OPPOSITE DIRECTION. THE TWIN ROTATED, TURNED TO HIS L. I WAS ABOUT 50 FT ALT ALSO THEN TURNED SLIGHTLY TO L, AND WE PASSED. THE TWIN COULD HAVE BEEN A C421 OR C414—REALLY DON'T KNOW. MAYBE TOTALLY DIFFERENT, BUT A FAIRLY LARGE TWIN. I HEARD NOTHING FROM THE ACFT AFTER TKO. I ASSUME THE TWIN HAD A BAD XMITTER, BECAUSE THERE WAS NO OTHER TFC IN THE PATTERN OR IN VICINITY. PRIOR TO TKO, THE ONLY XMISSIONS I HEARD WERE THE CARRIER SIGNALS.

SYNOPSIS 2 ACFT PERFORM A SIMULTANEOUS TKO FROM UDD, CA. THE PROB BEING, IN OPPOSITE DIRECTIONS AT THE SAME TIME.

Preparation and Remarks

The reporting pilot was prepared for takeoff from the airport. Appropriate radio transmissions were made on the proper frequency announcing the pilot's intentions—as required when operating from a nontowered airport. The opposing traffic may have had a faulty radio.

Postincident Analysis

This incident ended with no damage, but had both pilots not taken evasive action, they possibly would have struck one another. The selection of a runway is up to the pilot at uncontrolled (nontowered) airports. In this specific incident, two different pilots in two different aircraft selected the opposite ends of the same runway for takeoff.

The reporting pilot states that correct procedure was used in announcing location and intentions prior to take off. No other significant transmissions were heard, and takeoff proceeded as planned. Just as the aircraft was breaking ground, another aircraft appeared—also taking off—from the opposite direction. No mention is made in the report if the landing lights were switched on for either aircraft.

Each aircraft took evasive measures, and a collision was avoided. The evasive measures, although performed incorrectly, were effective.

The classification for this incident is pilot deviation, and the runway incursion severity category is category A because separation was decreased and participants had to take extreme action to narrowly avoid collision by turning away from each other.

The scenario shows four distinct problems.

Problem 1. There is no determination for runway assignment at an uncontrolled airport. The choice of runway usage is at the discretion of the pilot based on wind conditions and other factors.

Problem 2. All aircraft operating from nontowered airports must have properly functioning radios to make announcements with and to hear announcements from other aircraft using the same airport. Monitoring these announcements allows the pilot to visualize where other air traffic is and what they are doing. However, in this case, there was nothing to monitor except for a carrier signal.

Problem 3. Not all aircraft at uncontrolled airports have radio equipment (or working radio equipment); therefore, you must be visually vigilant.

Problem 4. Both aircraft turned left after rotation. Assuming that this was done as a collision avoidance maneuver, it was incorrect. Federal Aviation Regulations (FAR), section 91.113(e), says, "Approaching head-on. When aircraft are approaching each other head-on, or nearly so, each pilot of each aircraft shall alter course to the right." In this reported scenario, had one pilot done what was proper, they would have met. Perhaps, two wrongs sometimes do make a right!

Lessons Learned

The lesson learned from this report is to expect the unexpected at uncontrolled airports. The only indication of other traffic was an unmodulated carrier (radio signal with no voice).

Lesson 1. Expect the unexpected, and keep heads up. Always look for active traffic at any nontowered (uncontrolled) airport—in the air and on the ground.

Lesson 2. Turn on your landing lights, if so equipped, when landing and taking off from an uncontrolled airport.

Lesson 3. FAR 91.113(e) says, "Approaching head-on. When aircraft are approaching each other head-on, or nearly so, each pilot of each aircraft shall alter course to the right."

CASE 2

Landed at a Closed Airport

ASRS accession number: 456080

Month and year: November 1999

Local time of day: 0601 to 1200

Facility: 5T5, Hillsboro Municipal Airport (FIG. 5-2)

Location: Hillsboro, TX

Flight conditions: VMC

Aircraft 1: PA-24 Piper Comanche

Pilot of aircraft 1: Single pilot, 15,000 hours

Reported by: Pilot

Incident description: Runway incursion to closed runway

Incident consequence: None

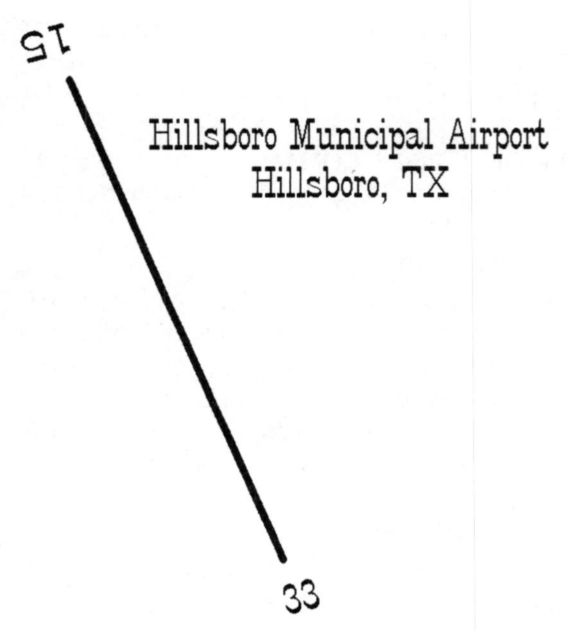

5-2 *Hillsboro Municipal Airport (this airport diagram is not suitable for navigational purposes).*

NARRATIVE LANDED AT HILLSBORO, TX, ARPT WHEN IT WAS NOTAM'ED CLOSED. FILED IFR FROM PANAMA CITY, FL (PFN) TO ALEXANDRIA, LA (AEX). CANCELED IFR AND PROCEEDED TO HILLSBORO (5T5) TO REFUEL. I DID NOT CHK WITH FLT SVC FOR NOTAMS FOR 5T5. ARPT WAS CLOSED FOR RESURFACING OF RWY. WORK HAD NOT BEGUN YET. NO BARRIERS IN PLACE ON RWY. NO CONFLICT OCCURRED. THE WAY TO PREVENT THIS IS TO ALWAYS CHK FOR NOTAMS WHEREVER A LNDG IS MADE. I DON'T THINK PEOPLE WILL ALWAYS DO THIS, THOUGH I DID MAKE ABOUT 5 RADIO CALLS ON UNICOM GOING INTO 5T5. MAYBE SOMEONE SHOULD MONITOR THE FREQ WHEN THE ARPT IS CLOSED. AFTER I FOUND THE ARPT WAS CLOSED, I ASKED A TEXAS DOT SUPVR ON SITE

IF THERE WAS ANY PROB AND IF WE COULD DEPART. HE REPLIED THERE WAS NO PROB BECAUSE THEY HADN'T STARTED WORKING YET, AND THAT IT WAS OK FOR ME TO TAKE OFF.

SYNOPSIS PLT OF A PIPER COMANCHE 250 UNKNOWINGLY LANDED ON A CLOSED ARPT FOR REFUELING.

Preparation and Remarks

The pilot was not prepared for the flight. Admission is made of not checking for NOTAMs when IFR plan was filed "I DID NOT CHK WITH FLT SVC FOR NOTAMS FOR 5T5."

The proper radio procedure for a nontowered airport was done, but no one was listening. The temporarily closed airport was not marked as required.

Postincident Analysis

This incident was caused by the pilot's failure to check NOTAMs for the intended destination. It was further exacerbated by the lack of markings on the airport runway signifying that it was closed. This incident is a pilot deviation, but there was no damage done to anyone or anything.

Three problems from the incident are noted:

Problem 1. The NOTAMs were not checked prior to making the flight. AIM suggests in Section 5-1-2 a(1), "Obtain a complete preflight and weather briefing. Check the NOTAMs."

Problem 2. The runway was not marked with a large yellow X at each end to indicate temporary closure.

Problem 3. This could have been more serious if construction equipment had been on the runway or digging had already been done on the runway—and the runway was not marked closed.

Although there is not a classification that includes airport management as being with fault, in this case there is at least some fault with airport management for not clearly marking the runway as closed.

Lessons Learned

This incident provides lessons for both the pilot and the airport operator. Neither had been prepared for the incident.

Lesson 1. Always check NOTAMs for information relative to airports you are operating from, going to, and passing over (along the flight route).

Lesson 2. Airport operators must physically mark runways that are closed for any reason. See AIM 2-3-6e, which requires a visual indication to pilots that a runway is temporarily closed in the form of yellow crosses at each end of the runway.

CASE 3

Opposite Ends of the Runway

ASRS accession number: 459588

Month and year: January 1999

Local time of day: 1201 To 1800

Facility: HYI, San Marcos Municipal Airport (FIG. 5-3)

Location: San Marcos, TX

Flight conditions: VMC

Aircraft 1: SA-277 (in experimental category)

Aircraft 2: Beech T-34

Pilot of aircraft 1: Captain, 5100 hours; first officer

Pilot of aircraft 2: Single pilot

Reported by: Pilot of SA-277

Incident description: Runway incursion

Incident consequence: Aborted takeoff

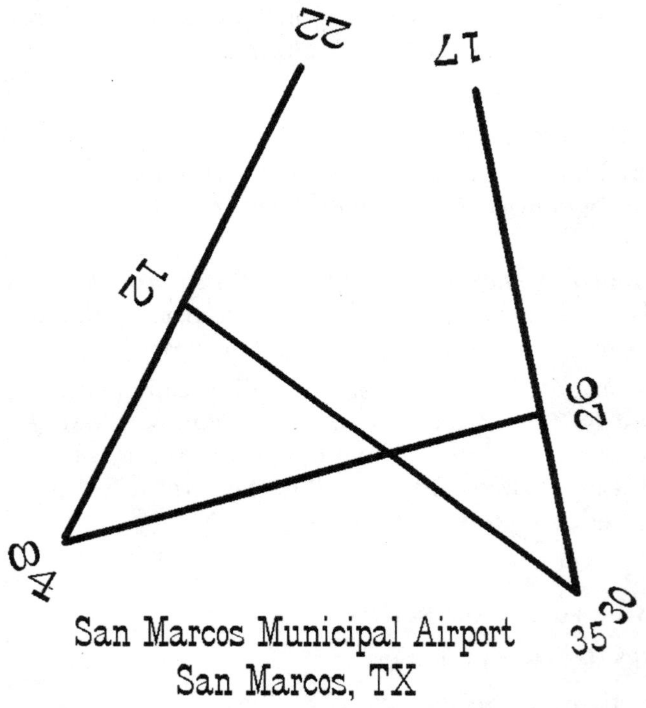

5-3 *San Marcos Municipal Airport (this airport diagram is not suitable for navigational purposes).*

NARRATIVE ON JAN/XA/00 AT SAN MARCOS, TX, I WAS ENGAGED IN THE FLT TESTING OF A NEW STC FOR THE INSTALLATION OF BRAKES MANUFACTURED BY XYZ ON AN SA-227 WHILE THE ACFT WAS TEMPORARILY IN THE EXPERIMENTAL CATEGORY. TESTING PARAMETERS REQUIRED THAT THE ACFT BE FLOWN ON RWY 12 ILS TO 50 FT ABOVE THE RWY THRESHOLD, LANDED AND STOPPED WITHIN A SPECIFIC DISTANCE. SAN MARCOS IS AN UNCONTROLLED FIELD AND I BROADCASTED ALL ACFT MOVEMENTS BOTH ON THE GND AND IN THE AIR ON THE CTAF FREQ (123.05). PRIOR TO START WE HAD CALCULATED AND CAME TO THE CONCLUSION THAT WE

WOULD UTILIZE RWY 26 FOR TKO. I TAXIED DOWN TO RWY 35 EN ROUTE TO RWY 26 AND STATED MY ACTIONS ON CTAF. I STOPPED SHORT OF RWY 26 AND PERFORMED ALL NECESSARY CHECKS. I TAXIED INTO POS AND STATED MY TKO AND INTENTIONS ON CTAF. A VISUAL CHECK INDICATED A CLR RWY AND I STARTED MY TKO ROLL. FULL PWR WAS IMMEDIATELY UTILIZED AS TKO WAS NEAR MGTOW. AT APPROX 100 KTS AND ABOUT 2000 FT INTO MY TKO ROLL I SAW ANOTHER ACFT ON OPPOSITE END OF THE RWY, RAPIDLY INCREASING IN SIZE. I IMMEDIATELY PULLED THE ENGS INTO FULL REVERSE AND APPLIED MAX BRAKING. I WAS MOVING SLOWLY AND NEARLY AT A FULL STOP WHEN THE OTHER ACFT, A T34, FLEW OVERHEAD AT VERY CLOSE RANGE. I CONTACTED THE T34 AND ASKED HIM WHY HE DID NOT BROADCAST HIS INTENTIONS. HE RESPONDED THAT HE DID AND ASKED ME WHY I HAD NOT BROADCAST MINE, AS HE DID NOT HEAR MY XMISSIONS. I STATED THAT I DID, AND THAT I HAD WITNESSES TO THAT FACT. I WAS PROPERLY SET UP AND BROADCASTING ON FREQ. CONVERSATION WITH ONE OF THE PRINCIPALS AT THE COMPANY INDICATED THAT THEY COULD HEAR THE T34 ON THE END OF RWY 8 (THE PRINCIPAL HAD A HANDHELD RECEIVER). HE STATED THAT MY BROADCASTS BROKE UP AS I APPROACHED RWY 35 DURING TAXI, WHICH AFTER THAT POINT HE COULD NOT HEAR ANY FURTHER XMISSIONS FROM ME. INTERESTINGLY ENOUGH, AFTER TALKING TO TEST PERSONNEL THAT WERE LOCATED NEAR THE MIDPOINT OF THE RWY, THEY STATED THAT THEY COULD ONLY HEAR MY XMISSIONS AND THEY NEVER HEARD ANYTHING FROM THE T34. ONE OTHER CONTRIBUTING FACTOR IS THAT THERE IS SMALL RISE IN THE MIDDLE OF THE ARPT THAT OBSCURES THE VIEW OF THE OPPOSITE END OF THE RWY. THAT IS WHY I WAS NOT ABLE TO SEE THE OTHER ACFT UNTIL I WAS FURTHER DOWN THE RWY. THIS RISE MIGHT ALSO AFFECT THE ABILITY TO XMIT/RECEIVE ON THE CTAF WHILE ON THE GND. THERE SHOULD BE A NOTAM POSTED AT THIS FIELD INDICATING

RADIO XMISSIONS/RECEPTION ANOMALIES PRESENT AT THIS ARPT. ACFT MANAGEMENT MIGHT WANT TO CONSIDER THE INSTALLATION OF A LOW WATTAGE RADIO REPEATER AT THE CTR OF THE FIELD TO ALLEVIATE THIS PROB. CALLBACK CONVERSATION WITH THE RPTR REVEALED THE FOLLOWING INFO: RPTR FELT THAT SOMETHING SHOULD BE DONE ABOUT THE SIT BUT HADN'T KNOWN WHERE TO START.

SYNOPSIS AN SA227 ABORTS ITS TKO WHEN A T34 IS NOTED ON TKO ROLL, OPPOSITE DIRECTION, SAME RWY AT SAN MARCOS, HYI, TX. BOTH ACFT XMITTED TKO INTENTIONS, NEITHER ONE HEARD THE OTHER.

Preparation and Remarks

According to the report, the SA-227 was operating as would be expected. Announcements were being made on the correct frequency. Also, according to the report, witnesses say that both aircraft did transmit information as they should have. Neither aircraft reported hearing each other—until after the initial incident.

Postincident Analysis

The report shows either a distinct radio anomaly or radio equipment not working properly. Two aircraft—about a mile apart—were not able to hear each other. People in the approximate middle of the runway's length report hearing only one aircraft's transmissions. Another subject reports hearing the T-34's transmissions.

There is an indication in the report that there is a slight rise in the middle of the airport and that this rise might have affected radio transmissions. The slight rise between ends of the runway is not likely to interfere with radio communications in the VHF range unless there are underlying equipment malfunctions.

A rise of a very few feet, with no other obstacles in the way, would not prevent a VHF signal from being heard over the distance indicated—unless the radio transmitters were extremely low powered (less than a couple of watts) or the transmissions were being overridden by some unknown form of radiofrequency interference or blockage. VHF radio propagation generally is line of sight. If you can see from one point to another, you can radio communicate from those same points. In practice, a very slight visual interruption, as indicated in this report, would have little or no effect on VHF propagation.

The classification for this incursion would be pilot deviation, and the runway incursion severity category would be category A because the SA-227 had to take extreme evasive measures to stop the aircraft.

A significant concern not specifically noted in the report is why those personnel standing in the middle of the runway's length made no comment to either aircraft about an impending collision and why another witness took no action when that person had knowledge of the T-34 taking off on the same runway.

Several problems are noted:

Problem 1. It is not possible to visually check the full length of the runway from either end. Was it possible for those in the middle of the runway's length to see both aircraft?

Problem 2. The radio equipment on both aircraft should be checked because there is no reasonable explanation for a communications failure at that range.

Problem 3. Other personnel were monitoring the situation and should have seen what was about to happen—yet remained silent until after the fact—"CONVERSATION WITH ONE OF THE PRINCIPALS AT

THE COMPANY INDICATED THAT THEY COULD HEAR THE T-34 ON THE END OF RWY 8 (THE PRINCIPAL HAD A HANDHELD RECEIVER)."

Lessons Learned

The San Marcos airport is defective in that a small rise in the middle of the airport prevents pilots from seeing the far end of the runway—thus not allowing pilots a necessary visual check for traffic.

Lesson 1. If at all possible, use runways on which you can check for traffic visually. In this way, you will not be surprised by the unknown.

Lesson 2. Check your radios for proper operation.

CASE 4

The Squeeze Play

ASRS accession number: 132503

Month and year: January 1990

Local time of day: 1201 to 1800

Facility: P10, Scandia Air Park (FIG. 5-4)

Location: Russell, PA

Flight conditions: VMC

Aircraft 1: GA type small single

Aircraft 2: GA type unknown

Pilot of aircraft 1: Single Pilot

Pilot of aircraft 2: Unknown

Reported by: Pilot

Incident description: Runway incursion

Incident consequence: None

NARRATIVE THERE WERE SEVERAL LIGHT ACFT IN THE PATTERN, ALL USING THE CTAF. THE ACFT WERE CLOSELY

5-4 *Scandia Air Park (this airport diagram is not suitable for navigational purposes).*

SPACED. AS 1 ACFT CROSSED THE THRESHOLD, I ROLLED OUT ON THE RWY BEHIND HIM. ANOTHER ACFT TURNED FROM BASE TO FINAL. THERE WAS ADEQUATE DISTANCE BTWN ALL 3 ACFT. WHEN THE LNDG ACFT BEGAN HIS TURN OFF THE RWY, I BEGAN MY TKO ROLL, MINDFUL OF THE ACFT ON FINAL BEHIND ME. AS I ACCELERATED, THE ACFT IN FRONT OF ME STOPPED HIS TURN AND CONTINUED DOWN THE RWY. THERE WAS ENOUGH DISTANCE BTWN US FOR ME TO CONTINUE MY TKO. WHEN AIRBORNE, I MOVED TO THE LEFT SLIGHTLY TO AVOID FLYING OVER HIM. BY THE TIME I PASSED HIM HE WAS JUST CLRING THE RWY. I FELT THAT ABORTING THE TKO WOULD HAVE RESULTED IN MY SLIDING UP CLOSELY

BEHIND HIM, AND THE ACFT BEHIND ME WOULD HAVE COMPOUNDED THE PROB. HAD THE FIRST ACFT TURNED OFF WHEN AND WHERE HE FIRST BEGAN TO TURN, THERE WOULD NOT HAVE BEEN A PROB. WITH THE BENEFIT OF HINDSIGHT, IT'S CLEAR THAT I COULD HAVE PREVENTED THIS OCCURRENCE BY WAITING FOR A BIGGER GAP BTWN ACFT BEFORE BEGINNING MY TKO ROLL. AT THE TIME THERE REALLY WAS ENOUGH ROOM, BUT SEEING HOW IT ALL SOURED I WILL ALLOW A GREATER MARGIN FOR SAFETY IN THE FUTURE.

SYNOPSIS SMA TAKING OFF FROM UNCONTROLLED AIRPORT DEPARTS TOO QUICKLY BEHIND LNDG TRAFFIC.

Preparation and Remarks

The pilot reports using standard radio procedure for a nontowered airport operation and that visual observations of traffic were being made. When an opening in the traffic was seen, the pilot moved in and took the opening. The opening then changed—becoming smaller than planned.

Postincident Analysis

The pilot was operating the radio with no problems, making the required announcements and monitoring other radio and visual traffic at the airport. No problems noted for communications. The pilot entered the takeoff line between landing airplanes—requiring very good timing and good luck. Luck is not a dependable resource in aviation.

As the pilot noted in the report, "WITH THE BENEFIT OF HINDSIGHT, IT'S CLEAR THAT I COULD HAVE PREVENTED THIS OCCURRENCE BY WAITING FOR A BIGGER GAP BTWN ACFT BEFORE BEGINNING MY TKO ROLL. AT THE TIME THERE REALLY WAS ENOUGH ROOM, BUT SEEING HOW IT ALL SOURED I WILL ALLOW A GREATER MARGIN FOR SAFETY IN THE FUTURE." The indication is that a slight delay in takeoff, as in waiting for all other aircraft to land, would have been prudent.

Three distinct problems are noted:

Problem 1. The pilot was in too much of a hurry to take off—which is claimed later as "WILL NOT HAPPEN IN THE FUTURE."

Problem 2. The runway is only 2000 ft long, making the timing for getting between other moving aircraft a serious gamble.

Problem 3. The pilot did not have a plan for use in the event something went wrong.

Lessons Learned

It does not pay to do anything in a hurry. It is kind of like the old adage, "The more hurried I get, the farther behind I am." This entire incident could have been avoided if the pilot had been somewhat more reserved in choosing the takeoff opening in the line of aircraft landing—as admitted in the report.

Lesson 1. Take your time. Do nothing in a hurry.

Lesson 2. Plan ahead: What will you do if .... For example, what would you have done in this scenario if the airplane landing in front of you had just plain stopped on the runway? Headed for the infield? Hope you didn't get eaten alive from the rear? The only sure answer was to have allowed more space between aircraft.

Lesson 3. Landing aircraft have the right of way [FAR 91.113(g)].

CASE 5

Buzz Job

ASRS accession number: 83639

Month and year: March 1988

Local time of day: 1201 to 1800

Facility: TRM, Desert Resort Regional Airport (FIG. 5-5)

Location: Palm Springs, CA

Flight conditions: VMC

Aircraft 1: SMT (small transport)

Aircraft 2: GA type centerline twin

Pilot of aircraft 1: PIC

Pilot of aircraft 2: Single pilot

Reported by: Pilot of aircraft 1

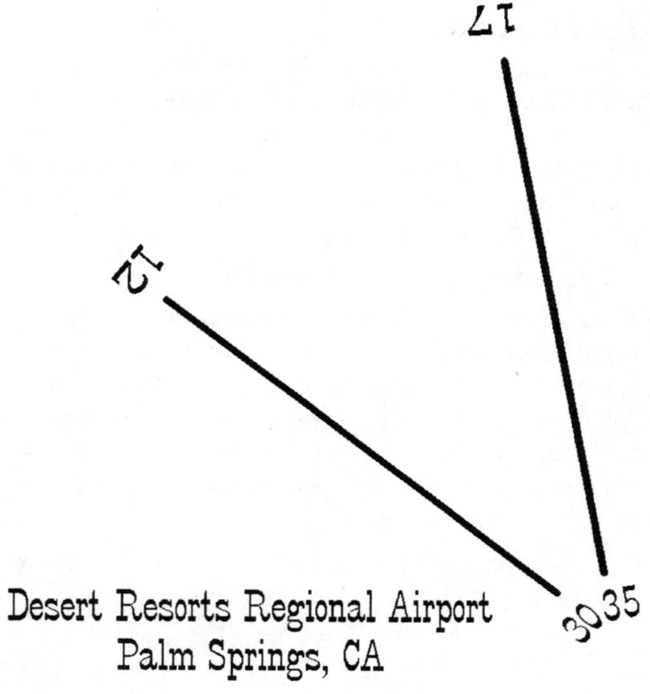

5-5 *Desert Resorts Regional Airport (this airport diagram is not suitable for navigational purposes).*

Incident description: Runway incursion

Incident consequence: Go around

NARRATIVE I STARTED THE ENGINES OF MY SMT. ON TAXI OUT I RADIOED MY INTENTIONS TO TAXI TO AND DEPART RWY 17 AT TRM. THE THERMAL FLIGHT SERVICE STATION CALLED BACK SAYING RWY IN USE WAS 12. I THEN PROCEEDED TO RWY 12. DURING THE TAXI THE FSS STATION ADVISED THAT THERE WAS TFC IN THE PATTERN. ALL THIS TIME I WAS LISTENING TO WHERE ALL THE TFC WAS IN REFERENCE TO TRM AIRPORT. DUE TO THE AREA OF TAXIWAYS IT IS POSSIBLE TO SEE THE ENTIRE PATTERN OF 12 WHILE TAXIING OUT. I CHECKED FINAL: SAW NO ACFT. I CALLED PRIOR TO THE HOLDING LINE THAT I WAS TAKING THE ACTIVE 12 FOR A RIGHT TURN OUT. HALFWAY DOWN THE RWY A CENTERLINE THRUST TWIN ENGINE SMA OVER TAKES ME DIRECTLY OVERHEAD. I LIFTED OFF AND KEPT LOW NOT KNOWING WHERE THE SMA WENT. IT IS MY OPINION THAT THE SMA DID NOT CALL TURNING FINAL AND MADE A SHORT APCH. IN NO WAY DID HE TRY TO REMEDY THE SITUATION BUT INSTEAD COMPLICATED IT BY THE NEAR PASS. WHILE FLYING IN UNCONTROLLED AIRPORTS I FEEL THAT THERE SHOULD BE MORE DILIGENCE IN THE LOCATION OF OTHER ACFT NOT ONLY VISUALLY BUT ALSO ON THE RADIO AND SHOULD A CONFLICT OCCUR THEN EACH PILOT SHOULD TAKE THE NECESSARY ACTION WHERE THE LEAST RISK OF AN ACCIDENT WOULD OCCUR. CALLBACK CONVERSATION WITH RPTR REVEALED THE FOLLOWING: SMA ACFT WAS A TRAINING FLIGHT AND THE REPORTER FEELS HE WAS NOT USING PROPER RADIO PROC. HE HAD CONTACTED THE ACFT AND ADVISED THEM TO CALL ON AN 800 NUMBER TO DISCUSS THE INCIDENT NEVER CALLED. HE NOTED AFTER THE CONVERSATION THE SMA DID NOT ADVISE HE WAS REMAINING IN THE PATTERN AND NEVER ANNOUNCED HIS POSITION. WHEN HE TURNED FINAL HE ADVISED BUT DID NOT GIVE THE ARPT OR RWY. UNDER THE

CONDITIONS HE ADMITTED HE COULD HAVE BEEN MORE VIGILANT AND LOOKED A LITTLE MORE CAREFULLY FOR TFC.

SYNOPSIS UNCONTROLLED ARPT ACFT ON FINAL DID NOT REPORT POSITION. FORCED TO GAR AS ACFT TAKING OFF.

Preparation and Remarks

The reporting pilot, flying a small transport aircraft, was prepared for a departure. Proper radio procedure reportedly was used, and a visual check was made prior to actually starting the takeoff. Note the statement in the report saying, "DUE TO THE AREA OF TAXIWAYS, IT IS POSSIBLE TO SEE THE ENTIRE PATTERN OF 12 WHILE TAXIING OUT. I CHECKED FINAL: SAW NO ACFT."

Postincident Analysis

This incident resulted in a loss of separation, but no accident resulted, and both aircraft involved continued their respective flights. The transport pilot made the required radio announcements to alert other aircraft in the vicinity to his intentions. The pilot also checked the area visually and saw nothing. Remember, a visual check cannot be made of a full 360 degrees unless the aircraft is turned—which was not stated in the report. The statement made on the report, "DUE TO THE AREA OF TAXIWAYS, IT IS POSSIBLE TO SEE THE ENTIRE PATTERN OF 12 WHILE TAXIING OUT. I CHECKED FINAL: SAW NO ACFT," appears to be incorrect.

At the Desert Sands Regional Airport, the terminal and FBO area are located between the ends of runways 12 and 17. This means that an aircraft taxiing toward runway 12 most likely would not see an aircraft in a standard traffic pattern (on downwind) approaching runway 12—unless it was on final at the time. To make a full scan for traffic prior to taking the active runway, the aircraft would have to make a 360-degree turn to the right. There is no indication in the report of this being done.

The overtaking centerline twin aircraft was wrong for buzzing the aircraft on the runway. The buzzing did nothing to enhance safety but perhaps made the pilot feel better. It was not justified and was very dangerous.

Several problems are noted from the incident.

Problem 1. The reporting pilot did not see the landing aircraft, and there is no indication of an attempt to see the full traffic pattern.

Problem 2. Not all pilots fulfill the requirements of radio usage at uncontrolled airports—hence the need for visual traffic checks.

Lessons Learned

Flight operations at nontowered airports require intensive visual vigilance on the part of the pilot(s) and require that proper radio announcements be made and listened to.

The lessons learned from this incident apply to all nontowered airport operations and are from AIM Section 4-1-9 (Traffic Advisory Practices at Airports Without Operating Control Towers).

Lesson 1. "There is no substitute for alertness while in the vicinity of an airport. It is essential that pilots be alert and look for other traffic and exchange traffic information when approaching or departing an airport without an operating control tower. This is of particular importance since other aircraft may not have communication capability or, in some cases, pilots may not communicate their presence or intentions when operating into or out of such airports. To achieve the greatest degree of safety, it is essential that all radio-equipped aircraft transmit/receive on a common frequency identified for the purpose of airport advisories."

Lesson 2. "Pilots of inbound traffic should monitor and communicate as appropriate on the designated CTAF from 10 miles to landing. Pilots of departing aircraft should monitor/communicate on the appropriate frequency from startup, during taxi, and until 10 miles from the airport unless the CFRs or local procedures require otherwise."

Lesson 3. FAR 91.113(f): "Overtaking. Each aircraft that is being overtaken has the right-of-way, and each pilot of an overtaking aircraft shall alter course to the right to pass well clear."

CASE 6

Deer Strike

ASRS accession number: 103546

Month and year: January 1989

Local time of day: 1801 to 2400

Facility: LND, Hunt Field (FIG. 5-6)

Location: Lander, WY

Flight conditions: VMC

Aircraft 1: Small transport (air taxi)

Pilot of aircraft 1: PIC, 5460 hours

Reported by: Pilot

Incident description: Incident (by NTSB)

Incident consequence: Substantial aircraft structural damage

NARRATIVE THE FOLLOWING HAS BEEN CLASSIFIED AS AN INCIDENT BY THE NTSB. IN 1/89 ON TKO ROLL FROM LND AT ROTATION SPD, APPROX 95 KTS, A RWY INCURSION AND STRIKE OCCURRED INVOLVING 2 DEER, A DOE AND A FAWN, IN TRAIL, XING THE RWY AT A RUN FROM S

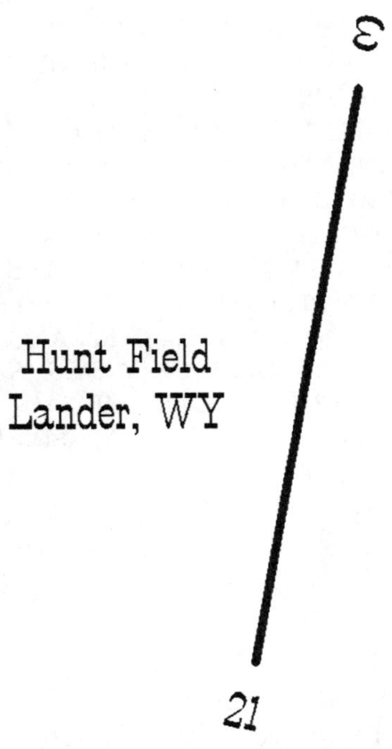

5-6 *Hunt Field (this airport diagram is not suitable for navigational purposes).*

TO N. AN EVASIVE ROTATION MANEUVER WAS ATTEMPTED IN AN EFFORT TO CLR THE DEER WHICH WAS PARTIALLY SUCCESSFUL REGARDING THE NOSE LNDG GEAR. ALTHOUGH ONLY ONE MODERATE STRIKE WAS HEARD ON THE BELLY OF THE ACFT RESULTING IN NO OBVIOUS CHANGE IN ENG OR ACFT PERFORMANCE, IT WAS DETERMINED AFTER THE FACT THAT BOTH DEER HAD MADE CONTACT SEVERING THE LEFT TIRE, LEFT BRAKE AND LEFT PISTON FROM THE ACFT AND FORCING THE LOWER LEFT STRUT ASSEMBLY TO BE AT A 45 DEG ANGLE TO THE REAR. MINOR DAMAGE WAS ALSO RECEIVED BY THE LEADING EDGE OF THE LEFT HORIZ STABILIZER. AFTER DETERMINING THAT THE LEFT TIRE, BRAKE AND PISTON

WERE NO LONGER ATTACHED TO THE ACFT, I ELECTED TO DO A FULL GEAR UP LNDG AT CASPER AND THE REMAINING GEAR WAS RETRACTED. AFTER PREPARING THE ACFT AND PAX FOR THE FULL GEAR UP LNDG THE ACFT WAS CONFIGURED AND ESTABLISHED ON A LONG STRAIGHT IN FINAL TO RWY 21. THE SUCCESSFUL GEAR UP LNDG WAS COMPLETED AT CASPER. ALL PERSONS ON BOARD THE ACFT WERE EVACUATED IN APPROX 40 SECS W/O INCIDENT OR INJURY. THE RESULTING ACFT DAMAGE FROM THE FULL GEAR UP LNDG WAS LIMITED AND TYPICAL OF THAT FOUND IN MOST GEAR UP LNDGS IN THIS TYPE OF ACFT. IT WAS LEARNED AFTER THE FACT FROM THE ARPT MGR THAT LND HAS HAD EXTENSIVE PROBS WITH GAME AND LIVESTOCK ON THE RWY YET A FENCE HAS NEVER BEEN INSTALLED.

SYNOPSIS ATX SMT HIT 2 DEER ON TKO, DAMAGING THE LEFT GEAR.

Preparation and Remarks

The pilot reports a normal takeoff run until a deer strike occurred, causing substantial damage to the aircraft—requiring a gear-up landing at another airport. Note that although animal strikes are more common at nontowered airports, they can and do happen at all airports. The pilot probably avoided a more serious incident by making an early takeoff rather than risking a full-impact collision with both animals or leaving the runway at rotation speed.

Postincident Analysis

This incident raises questions about pilot visual vigilance limitations and about the responsibility of airport management. During the hours of darkness, it is extremely difficult to see what may be running across an airport. The only light you may be working with comes from the landing lights on the aircraft, which provide only a narrow beam of lighted area to observe. Animal control at airports is difficult and expensive. Deer in particular are a fencing problem because they are excellent jumpers—quite able to clear an 8-ft fence with little effort.

The problems noted are:

Problem 1. There is considerable open distance surrounding the runway, and the deer should have been visible as they approached the runway. However, this incident occurred during the hours of darkness (time and date). Therefore, the deer only became visible when they were on the runway directly in front of the airplane.

Problem 2. There is no perimeter fence surrounding the airport to prevent animals from entering the runway area.

Problem 3. Airport management was aware of the animal problem on the runway, as stated in the report, "IT WAS LEARNED AFTER THE FACT FROM THE ARPT MGR THAT LND HAS HAD EXTENSIVE PROBS WITH GAME AND LIVESTOCK ON THE RWY, YET A FENCE HAS NEVER BEEN INSTALLED."

Lessons Learned

Expect the unexpected is the lesson learned from this report. There is little a pilot can do when an animal(s) runs from a dark area onto the runway.

Lesson 1. Keep your head out of the cockpit and be visually vigilant.

Lesson 2. Although expensive, proper fences can prevent animals from entering an airport or runway. However, the fences must be of a height designed to prevent jumping—which in the case of deer requires a 10-ft or greater height. Artificial lighting can aid in keeping animals out and certainly may make them more visible when they are present.

To some extent, other measures have been used to keep airports clear of animal intrusions. These include the use of control dogs, extensive local hunting, and audio devices.

CASE 7

Residential Airport

ASRS accession number: 168123

Month and year: January 1991

Local time of day: 0601 to 1200

Facility: Q68, Pine Mountain Lake Airport (FIG. 5-7)

Location: Groveland, CA

Flight conditions: VMC

Aircraft 1: GA type small single

Pilot of aircraft 1: Single pilot

Reported by: Pilot

Incident description: Runway incursion

Incident consequence: Evasive maneuver

NARRATIVE JUST AFTER T/D ON RWY 27 (Q68), A PICKUP TRUCK WAS OBSERVED APCHING THE RWY AND WAS ABOUT TO CROSS. THIS WOULD HAVE RESULTED IN A COLLISION WITH MY ACFT. I STILL HAD 35 KTS ON THE ROLL. TOO FAST TO STOP A TAIL DRAGGER, BUT

Pine Mountain Lake Airport
Groveland, CA

9 27

5-7 *Pine Mountain Lake Airport (this airport diagram is not suitable for navigational purposes).*

ENOUGH TO BALLOON OVER BY PULLING FULL FLAPS (I LANDED WITH NONE). FULL PWR BROUGHT ME BACK DOWN SAFELY W/O STALL/SPIN, BUT I DO BELIEVE I STALLED. VEH WAS GONE WHEN I LOOKED BACK AFTER ROLLOUT. BECAUSE THE ARPT IS 2980 MSL, MY SMA WAS LUCKY TO HAVE BEEN ABLE TO BALLOON OVER W/O STALL/SPIN. ANYWAY, THE PROB IS CLR. THERE ARE MANY ILLEGAL INCIDENTS OF VEHS, PEDESTRIANS, BICYCLES, SKATEBOARDS AND DOGS EVER DAY AT THIS AIRFIELD. THE RESIDENTS OF THIS ARPT COMMUNITY ARE THE PRIMARY OFFENDERS. WE NEED LAW ENFORCEMENT ACTION HERE IMMEDIATELY. I KNOW THIS IS NOT AN AIRSPACE SYS PROB, BUT THIS IS A PUBLIC ARPT WHICH IS SO CONSTRUCTED SO AS TO ALLOW VEHS EASY ACCESS W/O FEAR OF ENFORCEMENT ACTION. CALLBACK CONVERSATION WITH RPTR REVEALED THE FOLLOWING: TRUCK WAS MOVING FROM THE RAMP AREA TO THE N ACROSS THE RWY TO HOME ON THE N SIDE OF THE ARPT. ADVISED AN AB WAS BEING CONSIDERED, BUT NOT MUCH ELSE COULD BE DONE. FEELS THE PLT'S ASSN IS PROBABLY THE MAIN OFFENDERS. SAYS HE CLBED DIRECTLY OVER THE VEH. WOULD LIKE TO SEE SOME TICKETS ISSUED TO OFFENDERS. A CONTINUING PROB.

SYNOPSIS VEHICLE CROSSES IN FRONT OF LNDG ACFT. CLIMB OVER VEHICLE AND LAND.

Preparation and Remarks

The pilot was landing when a vehicle crossed the runway in front of the aircraft. An evasive (however, dangerous) maneuver was performed by the pilot—thus avoiding the vehicle. The airport where this incident happened is a residential field. In other words, there are homes, hangars, and privately owned vehicles on the airport. The scenario described is not unique to Q68 and is evident at many similar facilities.

Postincident Analysis

There was a loss of separation between the aircraft and the vehicle, causing the pilot to make an evasive maneuver. Had this aircraft been unable to make the evasive maneuver, there would have been a collision between the vehicle and the airplane. The classification for this incursion is a vehicle/pedestrian deviation, and the runway incursion severity category would be category A, just short of a collision with the pickup truck.

Several problems are raised with this situation:

Problem 1. Did the pilot maintain visual vigilance during landing? Was the truck visible before it appeared on the runway?

Problem 2. The vehicle should not have been on the active runway. Signage and an education program for residents about vehicle operation on the airport would be appropriate

Problem 3. Perimeter roadways should be provided for vehicles and no access allowed to the runway.

Problem 4. The airport is open to the public—perhaps too open.

Lessons Learned

Operations at residential airports sometimes can be quirky. Residents tend to believe that they have full and all-encompassing rights to do as they please. Most are pilots and feel that being a pilot gives them expertise around an airport. With this in mind, a pilot is well advised to expect nearly anything to happen. Residential airports have all the offerings of a residential suburban area, including errant vehicles, bicycles, children, pets, and flying toys (kites and model airplanes).

The lessons learned from this incident have to do with pilot visual vigilance and airport resident responsibility.

Lesson 1. The pilot must be visually vigilant of the surroundings—watching for movement on the airport toward the active runway. Vehicles, children, animals, and more can be found at residential airports—too often on and around the runway.

Lesson 2. Ground vehicle operators must be vigilant and yield the right of way to aircraft operating into, out of, and on the airport. Aircraft have the right of way on runways.

CASE 8

A Fatal Story

The following fatal incident, as taken from National Transportation Safety Board (NTSB) files, is a worst-case scenario of what can go wrong when two aircraft meet at intersecting runways, one landing and the other taking off.

NTSB report: AAR-97-04

Month and year: November 19, 1996

Local time of day: 1801 CST

Facility: UIN, Quincy Regional–Baldwin Field (FIG. 5-8)

Location: Quincy, IL

Flight conditions: VMC

Aircraft 1: Beech 1900C

Aircraft 2: Beech King Air

Aircraft 3: GA type single engine (Cherokee)

Pilot of aircraft 1: Captain, 4000 hours; first officer, 1950 hours

Pilot of aircraft 2: PIC, 25,647 hours

Pilot of aircraft 3: PIC

Incident description: Runway incursion

Incident consequence: Fatal collision

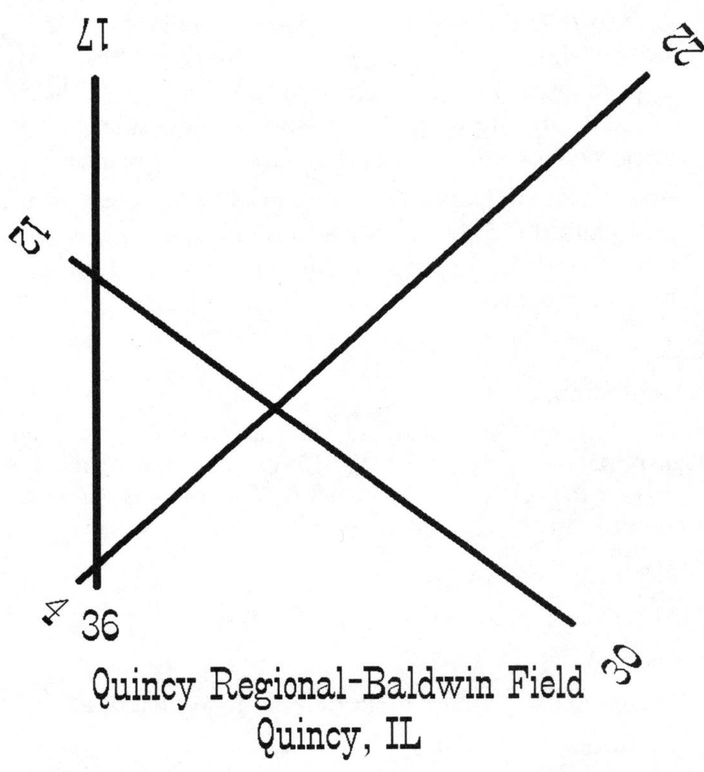

5-8 *Quincy Regional-Baldwin Field (this airport diagram is not suitable for navigational purposes).*

NARRATIVE ON NOVEMBER 19, 1996, AT 1701 CENTRAL STANDARD TIME, UNITED EXPRESS FLIGHT 5925, A BEECHCRAFT 1900C, N87GL, COLLIDED WITH A BEECHCRAFT KING AIR A90, N1127D, AT QUINCY MUNICIPAL AIRPORT, NEAR QUINCY, ILLINOIS.

FLIGHT 5925 WAS COMPLETING ITS LANDING ROLL ON RUNWAY 13, AND THE KING AIR WAS IN ITS TAKE OFF ROLL ON RUNWAY 04. THE COLLISION OCCURRED AT THE INTERSECTION OF THE TWO RUNWAYS. ALL 10 PASSENGERS AND TWO CREW MEMBERS ABOARD FLIGHT 5925

AND THE TWO OCCUPANTS ABOARD THE KING AIR WERE KILLED.

FLIGHT 5925 WAS A SCHEDULED PASSENGER FLIGHT OPERATING UNDER THE PROVISIONS OF TITLE 14 *CODE OF FEDERAL REGULATIONS* (CFR) PART 135. THE FLIGHT WAS OPERATED BY GREAT LAKES AVIATION, LTD., DOING BUSINESS AS UNITED EXPRESS. THE KING AIR WAS OPERATING UNDER 14 CFR, PART 91.

ACCORDING TO THE COCKPIT VOICE RECORDER (CVR), AT 1652:07, A FEMALE VOICE IDENTIFIED AS THE CAPTAIN OF FLIGHT 5925 STATED ON THE QUINCY COMMON TRAFFIC ADVISORY FREQUENCY (CTAF) THAT THE AIRPLANE WAS ABOUT 30 MILES NORTH OF THE AIRPORT AND THAT THEY WOULD BE LANDING ON RUNWAY 13; SHE ALSO ASKED THAT "ANY TRAFFIC IN THE AREA PLEASE ADVISE." NO REPLIES WERE RECEIVED TO THIS REQUEST.

AT 1655:19, A FEMALE VOICE, IDENTIFIED AS ONE OF THE OCCUPANTS IN THE KING AIR, ANNOUNCED, "QUINCY TRAFFIC, KING AIR ONE ONE TWO SEVEN DELTA'S TAXIING OUT...TAKE OFF ON RUNWAY FOUR, QUINCY."

AT 1655:40, THE CVR RECORDED THE VOICE OF THE MALE PILOT OF A PIPER CHEROKEE, WHICH WAS TAXIING BEHIND THE KING AIR, ANNOUNCING, "QUINCY TRAFFIC, CHEROKEE SEVEN SIX FOUR SIX JULIET BACK-TAXI...TAXIING TO RUNWAY FOUR, QUINCY."

AT 1655:48, THE CAPTAIN OF FLIGHT 5925 COMMENTED TO THE FIRST OFFICER, "THEY'RE BOTH USING [RUNWAY] FOUR." THE CAPTAIN THEN ASKED, "YOU'RE PLANNING ON ONE THREE STILL, RIGHT?" THE FIRST OFFICER REPLIED, "YEAH, UNLESS IT DOESN'T LOOK GOOD THEN WE'LL JUST DO A DOWNWIND FOR FOUR BUT...RIGHT NOW PLAN ONE THREE."

AT 1656:56, THE CAPTAIN OF FLIGHT 5925 ANNOUNCED OVER THE CTAF, "QUINCY AREA TRAFFIC, LAKES AIR TWO FIFTY ONE IS A BEECH AIRLINER CURRENTLY TEN MILES TO THE NORTH OF THE FIELD. WE'LL BE INBOUND TO

ENTER ON A LEFT BASE FOR RUNWAY ONE THREE AT QUINCY ANY OTHER TRAFFIC PLEASE ADVISE." THERE WAS NO RESPONSE.

AT 1659:03, THE FEMALE OCCUPANT OF THE KING AIR ANNOUNCED, "QUINCY TRAFFIC, KING AIR ONE ONE TWO SEVEN DELTA HOLDING SHORT OF RUNWAY FOUR. BE...TAKIN' THE RUNWAY FOR DEPARTURE AND HEADING...SOUTHEAST, QUINCY."

AT 1659:19, THE CAPTAIN OF FLIGHT 5925 COMMENTED, "SHE'S TAKIN' RUNWAY FOUR RIGHT NOW?" THE FIRST OFFICER REPLIED, "YEAH."

ACCORDING TO THE CHEROKEE PILOT, THE KING AIR PULLED UP FAR ENOUGH ON RUNWAY 04 TO ALLOW THE CHEROKEE ACCESS TO RUNWAY 36, AND WHEN THE KING AIR WENT INTO POSITION ON RUNWAY 04, HE TAXIED THE CHEROKEE INTO THE RUNUP AREA SHORT OF THE RUNWAY.

Note 1. Company practice is for the pilot not flying to handle radio communications.

Note 2. CTAF is a radio frequency designated for use by pilots operating near uncontrolled airports. Pilots use this frequency to broadcast their positions or intended flight activities or ground operations.

Note 3. In a postaccident interview, the Cherokee pilot indicated that when he heard this broadcast, he saw the airplane. He also said that he remembered seeing lights on each wing of the airplane on its final.

At 1700:16, the captain of flight 5925 reported that the airplane was "ON SHORT FINAL FOR RUNWAY ONE THREE" and asked, "THE AIRCRAFT GONNA HOLD IN POSITION ON RUNWAY 4 OR YOU GUYS GONNA TAKE OFF?" The King Air did not respond to this request.

At 1700:28, the pilot of the Cherokee stated, "SEVEN SIX FOUR SIX JULIET...HOLDING...FOR DEPARTURE ON RUNWAY 4."

The CVR then recorded an interruption in this transmission by a mechanical "two hundred" alert announcement from the ground proximity warning system (GPWS) in the Beech 1900C. The CVR then recorded the last part of the transmission from the Cherokee as "... [unintelligible word] ON THE UH, KING AIR."

At 1700:37, the captain of flight 5925 replied, "OK, WE'LL GET THROUGH YOUR INTERSECTION IN JUST A SECOND SIR [UNINTELLIGIBLE WORD] WE APPRECIATE THAT."

According to the Cherokee pilot and his passenger, as well as a pilot who saw the approach and landing of flight 5925 as he was driving to the airport, flight 5925 had its landing lights on. The passenger in the Cherokee said the airplane made a normal landing on runway 13. Time and distance data from an aircraft performance and visibility study conducted by the NTSB indicated that the King Air began its takeoff roll about 13 seconds before flight 5925 touched down on the runway at 1700:59. According to the occupants of the Cherokee, the King Air had been in position on the runway for about 1 minute before beginning the takeoff roll. The Cherokee pilot stated that he heard no takeoff announcement from the King Air over the CTAF, and none was recorded on the Beech 1900C CVR.

The Cherokee pilot indicated that he remembered hearing a call from the captain of flight 5925 while he was waiting in the runup area. However, he recalled her saying that the airplane was 2 miles out on final.

According to the CVR transcript, this was the Cherokee pilot's second radio call on the CTAF.

In a postaccident interview, the Cherokee pilot indicated that his recollection was that his second radio call before departing Quincy followed a transmission that included the words "KING AIR...TAXIING AND HOLDING." He stated that he did not hear the whole transmission but that at the time, he thought it came from the King Air, and he thought that the King Air pilot might have been talking to him. He said that after he asked his passenger whether he understood the transmission and his passenger said that he did not, he transmitted "KING AIR, THIS IS CHEROKEE 7646J…I AM RIGHT BEHIND YOU AND AM HOLDING FOR YOUR DEPARTURE."

The Cherokee pilot indicated that he did not recall hearing any transmission concerning an intersection.

A pilot waiting inside the FBO said he thought he remembered hearing a male voice "stepping on" (transmitting at the same time as) a female voice shortly before the accident.

None of the other witnesses who were listening to the CTAF at the time reported hearing such a "stepped on" transmission.

At 1701, during flight 5925's landing rollout, the airplane collided with the King Air at the intersection of runways 13 and 04.

NTSB Determination

The NTSB determined that the probable cause of this accident was the failure of the pilots in the King Air A90 to effectively monitor the common traffic advisory frequency (CTAF) or to properly visually scan for traffic, resulting in their commencing a takeoff roll when the Beech 1900C (United Express flight 5925) was landing on an intersecting runway. Contributing to the cause of the accident was the Cherokee pilot's interrupted radio transmission, which led to the Beech 1900C pilots' misunderstanding of the transmission as an indication from the King Air that it would not take off until after flight 5925 had cleared the runway.

REACT

As a result of this accident and at the request of the Federal Aviation Administration (FAA), which asked for help in reinforcing pilot education, the Air Safety Foundation of the Aircraft Owners and Pilots Association (AOPA) distributed a poster to more than 5000 FBOs and flight schools about uncontrolled airport operations. The poster includes the message, "Don't let a collision ruin your day—REACT."

REACT is an acronym for *r*adio, *e*yes, *a*nnounce, *c*ourtesy, and *t*raffic pattern. The poster also includes this list:

Radio. Listen for traffic and form a mental picture of the position and movement of all aircraft.

Eyes. Look for all traffic—including no-radio aircraft—and turn on landing lights when within 10 miles of the airport to make it easier to be seen.

Announce. Broadcast position on taking the active runway and on pattern entry, crosswind, downwind, base, and final.

Courtesy. Be courteous in the pattern. Discuss conflicts on the ground, not on the radio.

Traffic Pattern. Fly a standard traffic pattern at the recommended altitude.

In addition, the foundation published a "Safety Advisor" booklet on uncontrolled airport operations and published the pamphlet, "Pilot Operations at Nontowered Airports." You can reach the AOPA at *http://www.aopa.org* on the Internet.

Lessons Learned in this Chapter

This chapter has closely examined a number of different nontowered airport type runway incursions—as reported on the included ASRS reports used for the case studies. Pilot deviations and vehicle/pedestrian deviations were shown. The lesson is to recognize that aircrews can do little about vehicle/pedestrian deviation except for vigilance and reporting these deviations.

In each case study, several lessons were learned. Of these lessons, the following appear to be the most common:

Visual vigilance. Keep heads out of cockpits. Look around. Watch for other aircraft activity on the airport. Watch for any other activity on the airport.

Radio vigilance. By listening to and announcing your intentions on the proper CTAF you will know what is going on around you at all times and advise other airport users of your activities.

Airport diagrams. Airport diagrams will aid you in visualizing your approach to the correct runway and help you find where you are going on the ground. Note, however, that airport diagrams are not available for all nontowered airports.

Be visible. Use your landing lights when landing and taking off.

To look and listen cannot be emphasized enough. You must be very aware of your surroundings when operating into and out of nontowered airports.

6

Miscellaneous Runway Incursions

The following runway incursions are more bizarre in substance than most that have been studied in the earlier chapters of this book. The incidents vary from animal incursions, to flying debris, to out-and-out road rage. They are interesting in that they exemplify more of the unknown aspects of flying than previous examples. The primary lesson to be learned from the following Aviation Safety Reporting System (ASRS) report case studies is to expect the unexpected.

CASE 1

Loose Dog

ASRS accession number: 210579

Month and year: May 1992

Local time of day: 1201 to 1800

Facility: OPF, Opa Locka Airport (FIG. 6-1)

Location: Miami, FL

Flight conditions: VMC

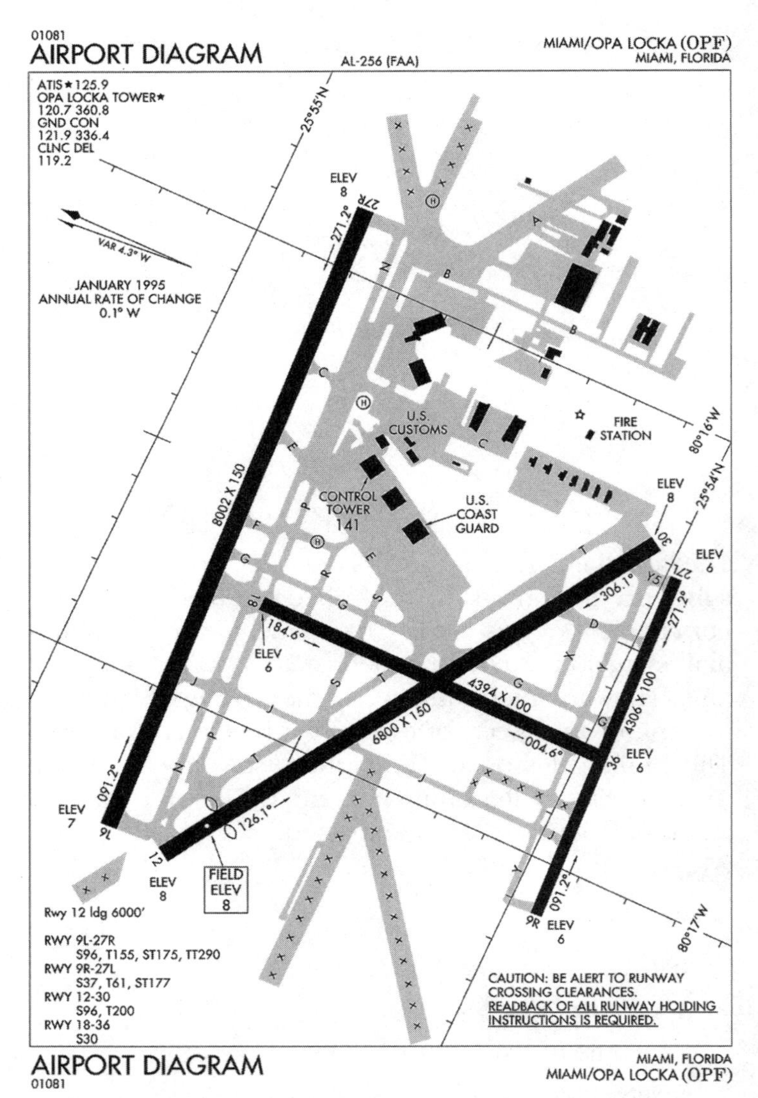

6-1 *Opa Locka, OPF (this airport diagram is not suitable for navigational purposes).*

Aircraft 1: GA type single

Pilot of aircraft 1: Single pilot, 2500 hours

Reported by: Pilot

Incident description: Runway incursion by a dog

Incident consequence: Aircraft structural damage

NARRATIVE DOG RAN ACROSS TAXIWAY FROM R TO L. HIT BRAKES HARD AND APPLIED FULL L RUDDER. MISSED DOG! L WING SCRAPED ACROSS METAL POST IMMEDIATELY TO L OF TAXIWAY. POST SERVES NO PURPOSE AND SHOULD BE REMOVED; SO SHOULD DOG BECAUSE THE NEXT TIME THIS HAPPENS TO HELL WITH IT, I'M GONNA HIT THE DOG ON THE THEORY DOG IS SOFTER THAN 6 INCH IRON PIPE FILLED WITH CONCRETE, AND SET IN CONCRETE, PLUS CHEAPER TO BURY DOG THAN FIX AIRPLANE!

SYNOPSIS SMA DODGES DOG ON TAXIWAY, HAS TAXIWAY EXCURSION AND HITS WING TIP ON CEMENT FILLED PIPE. ACFT DAMAGED. DOG MAKES GETAWAY.

Postincident Analysis

The pilot had to swerve hard to avoid a dog that was on the taxiway. In doing so, the aircraft was damaged when it struck a pipe protruding above ground level at the edge of the runway.

OPF is a reasonably large airport, and the pilot should have seen the dog coming from a distance. Visual range at the airport varies from several hundred to several thousand feet in most directions. Speed of the taxiing aircraft also may have been a factor.

This incident shows how a little laxness on the part of the pilot's visual vigilance can have expensive ramifications. Had the dog been spotted from a distance, there would have been no need for a hard evasive maneuver—which resulted in damage to the aircraft. Of course, had the dog been under the owner's control, the incident would have never happened.

Problems noted from this report include

Problem 1. The pilot did not see the dog coming.

Problem 2. The owner of (or person caring for) the dog should have been responsible for either confining or controlling the dog.

Problem 3. What is a cement-filled 6-inch pipe doing near a taxiway?

Lessons Learned

Vigilance, vigilance, vigilance—watch out for the unexpected. This lesson has been repeated throughout this book.

Lesson 1. Watch what is happening in your surroundings. It is much easier to slow in a straight line than to make an abrupt maneuver, with the possibility of losing control of the aircraft or otherwise damaging it.

Lesson 2. Avoid extreme maneuvers if possible, because they tend to force the aircraft into unusual attitudes—such as a wing very low in a sharp turn, allowing contact with the ground or objects on the ground.

CASE 2

No Place to Go

ASRS accession number: 314756

Month and year: August 1995

Local time of day: 1201 to 1800

Facility: CDK, George T. Lewis Airport (FIG. 6-2)

Location: Cedar Key, FL

Flight conditions: VMC

Aircraft 1: GA type single (Taylorcraft)

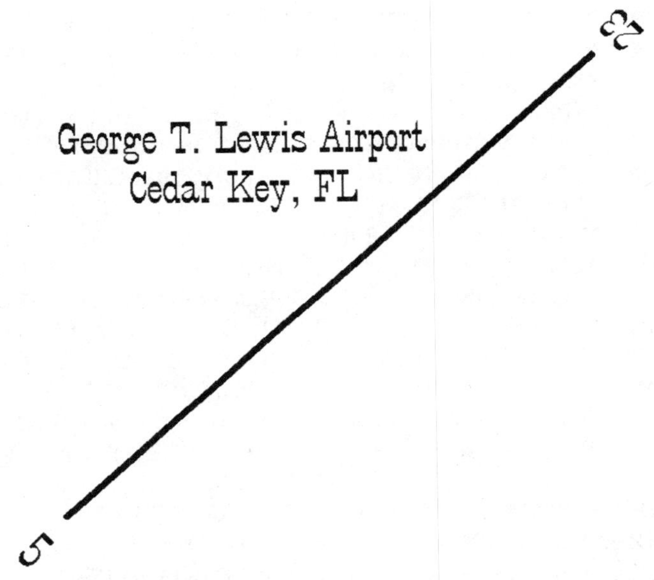

6-2 *George T. Lewis Airport (this airport diagram is not suitable for navigational purposes).*

Pilot of aircraft 1: Single pilot, 980 hours

Aircraft involved: GA type single

Reported by: Pilot

Incident description: Other

Incident consequence: Struck sign

NARRATIVE DURING BACK-TAXI FOR A DEP ON RWY 23 (FLT OF 4) ACFT SPOTTED ON SHORT FINAL FOR RWY 23. THE 3 REMAINING ACFT MADE R TURN ONTO GRASS STRIP BTWN RWY AND ROAD TO CLR RWY FOR LNDG TFC. THE LAST OR TRAIL ACFT ALSO TURNED OFF RWY TO THE R AND CONTACTED A SIGN. SAFETY CONCERNS MUST INCLUDE PLT AND PAX ALONG WITH AUTO AND PEDESTRIANS. (NOTE: NO TXWY EXISTS.) CONTRIBUTING FACTORS: 1) SIGN WAS POSTED VERY LOW, TOP OF SIGN

APPROX 40 INCHES (LOWER THAN DOT STANDARDS?) 2) SIGN WAS PLACED PARALLEL TO BOTH RWY AND ROAD. THIS MAKES SIGN VISIBLE FROM 90 DEGS THROUGH 45 DEGS, APPROX 3 FT WIDE, BUT THROUGH 180 DEGS SIGN IS ONLY APPROX 1/8-INCH WIDE, INSUFFICIENT. CORRECTIVE ACTION: SIGN SHOULD BE PLACED AT EYE LEVEL OR DOT STANDARD (REF STOP SIGN, ETC). PLACE SIGN ON OTHER SIDE OF ROAD. THIS PLACES SIGN OUT OF REACH OF BOTH ACFT AND AUTO. SIGN SHOULD ALSO BE (2) 90 DEG SIGNS PLACED AT 45 DEGS TO ROAD. CALLBACK CONVERSATION WITH RPTR REVEALED THE FOLLOWING INFO: RPTR STATES THAT ORIGINALLY THIS WAS A TXWY AND THE POWERS THAT BE JUST VOTED TO MAKE IT A ROAD FOR ACCESS TO HOUSES WHICH HAVE BEEN BUILT BEYOND THE ARPT. IT CAN NO LONGER BE USED FOR A TXWY. THERE ARE 4 SIGNS, AT THE ENTRY WAY, THE EXIT AND 2 PLACES IN BTWN. AS INDICATED THEY ARE NOT STANDARD AND TOO LOW FOR GOOD VISUAL CONTACT. THERE IS QUITE A BIT OF TFC IN THE AREA AS IT IS A SETTING FOR SEVERAL FESTIVALS DURING THE YR. THIS MAKES IT VERY INCONVENIENT FOR LNDG AND DEPARTING TFC COORD DUE TO HAVING TO BACK TAXI. THE LIGHTS HAVE BEEN CHANGED TO RECESSED LIGHTS DUE TO VANDALISM OF THE FRANGIBLE RWY LIGHTS. ANALYST SPOKE TO AN FAA ARPT CERTIFICATION SAFETY INSPECTOR REGARDING THE 'NOT FAA APPROVED' INDICATION IN THE ARPT FACILITY DIRECTORY. HE INDICATED IT IS BECAUSE OF THE LIGHTING, NOT THE ARPT ITSELF. IT REFS THE MIRL LIGHTING WHICH IS INDICATED AS NONSTANDARD. RPTR OF COURSE WOULD LIKE TO SEE STANDARDIZATION OF THE SIGNAGE TO COMPLY WITH FAA REGS, AND RELOCATION TO THE OTHER SIDE OF THE ROAD.

SYNOPSIS TAYLORCRAFT FORCED TO LEAVE RWY WHEN BACK-TAXIING DUE TO LNDG TFC HITTING SIGN IN GRASS AREA.

Postincident Analysis

The taxiing aircraft yielded the right of way to a landing aircraft. In so doing, a sign was struck. There is confusion at this airport about where airplanes and cars belong.

This airport offers little provision for taxiing aircraft. In this particular incident, the landing aircraft—seeing several airplanes taxiing on the runway—should have made the courteous choice of going around. This would have prevented a mass run from the runway for the taxiing airplanes. Of course, per the Federal Aviation Regulations (FARs), the landing aircraft did have the right of way.

This airport has been a victim of progress, with a housing area built in the vicinity of the airport and a taxiway converted to a road for access to those homes. Not mentioned in the report is the other problem of vehicular traffic too close to a runway—that of the occasional runway incursion by a vehicle.

The problems indicated by this report consist of poor signage, too small an escape area, and a conflict of vehicular and aircraft traffic:

Problem 1. Poorly planned and installed signage. The signs appear to have been placed as an afterthought, with little care or planning about placement or interest toward airport users.

Problem 2. Unplanned mixing of aircraft and vehicles. The signs are not going to keep the vehicles from mixing with airplanes at some point in time. A stout fence would.

Problem 3. The area the aircraft had to escape to appears to be narrow, since it was originally the strip between the taxiway and the runway. Again, the mark of poor planning.

Lessons Learned

Operations at very small airports often leave a lot to be desired in accommodations for arriving aircraft. They are more designed for regular users—those familiar with the airport's individual idiosyncracies.

Lesson 1. As small airport areas become residential, whether for aircraft owners or just for the building of more subdivisions, provisions must be made to keep aircraft traffic separated from vehicles, pedestrians, and animals.

Lesson 2. Always have a place to go when needed. In this case, a place to go was needed to get out of the way of the landing aircraft—which did have the right of way.

CASE 3

Open Access

ASRS accession number: 298962

Month and year: March 1995

Local time of day: 0601 to 1200

Facility: BVY, Beverly Municipal Airport (FIG. 6-3)

Location: Beverly, MA

Flight conditions: Various

Reported by: Airport user

Incident description: Runway incursion by pedestrian, etc.

Incident consequence: Not stated

NARRATIVE I HAVE SENT SEVERAL RPTS OVER THE PAST YR AND THEY ARE ALL DEALING WITH THE CONSTANT DEER PROB AND ENCROACHMENTS ON THE ARPT TXWYS AND RWYS. FOR SEVERAL MONTHS A HERD OF DEER HAVE BEEN ON THE ARPT SURFACE AND COULD HAVE CAUSED SEVERE PROBS. AN ACFT HAD TO BE SENT

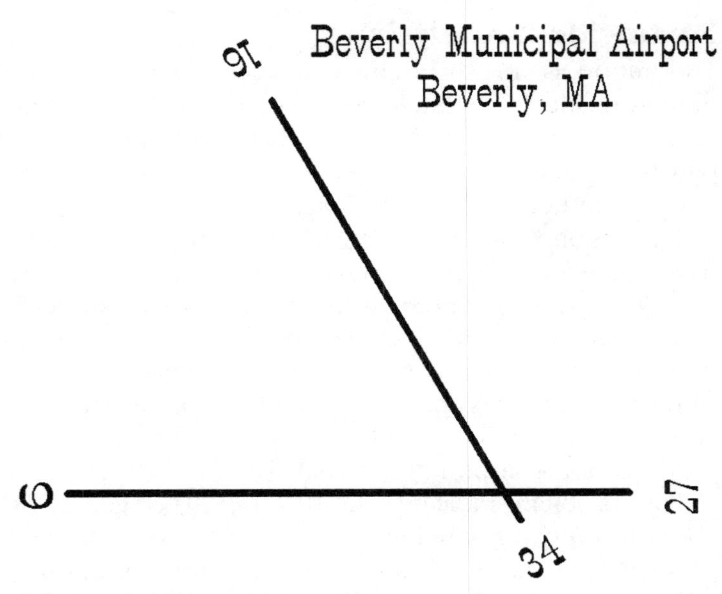

6-3 *Beverly Municipal Airport (this airport diagram is not suitable for navigational purposes).*

AROUND DUE TO DEER XING THE RWY PRIOR TO LNDG. THERE HAVE BEEN A COUPLE OF OTHER INCIDENTS AT THIS ARPT WHICH SHOULD BE BROUGHT TO THE ATTN OF THE PROPER AUTHS. A LCL POLICE VEHICLE ENTERED THE ARPT AND PROCEEDED DOWN TXWY B WITHOUT ANY COM WITH THE CTL TWR. 2 PEOPLE ON BIKES, WITH 2 DOGS, ENTERED THE ARPT SURFACE AT TXWY B AND CROSSED RWY 9, THE ACTIVE RWY. THERE WAS NO LOSS OF SEPARATION. SOMETHING HAS TO BE DONE ABOUT THESE SITS. VEHICLES AND PEOPLE CAN ENTER MANY PARTS OF THIS ARPT AND PROCEED ONTO MOVEMENT AREAS. A PERIMETER FENCE OR GATES WITH LOCKS SHOULD BE IN PLACE AT THE VARIOUS ENTRY POINTS.

SYNOPSIS RWY TXWY INCURSIONS BY DEER AND PEDESTRIANS.

Postincident Analysis

This report is not for a single incident, but rather for numerous incidents that have occurred over a period of time at the airport (BVY). Note that this airport has a part time control tower, generally operating during the daytime and early evening hours.

The report outlines several incidents that apparently have been on the reporting person's mind for some time. The primary concern is that the problem continues with no abatement in sight. The overall problem seems to be a lack of access control to the airport taxiways and runways. It appears that vehicles, bicycles, pedestrians, and pets have full and unlimited access to the active portions of the airport at all times.

The problem indicated in this report is generally about too much nonflying access and is very similar to problems encountered at many other airports around the country:

Problem 1. No gates of fences to keep unauthorized people, animals, and vehicles away from the active portions of the airport.

Lessons Learned

To prevent vehicle/pedestrian deviations (runway incursions) from happening at an airport, there must be some means of access control.

Lesson 1. Control entry onto the airport active areas. This may be in the form of signage directed toward the public stating the dangers and legal penalties for entering the active areas of the airport. More effective than signage are fencing and gates. Although some entries to the airport will still be made, most will be stopped by a fence and gate system. The most effective means of keeping the nonflying public from entering the active areas of an airport is a combination of signs and fences. Take a look at how military installations are protected as an example. Most potential airport encroachers generally are thwarted by the high fences and threatening signs.

Lesson 2. For those flying into and out of Beverly (and/or other similar airports), be very vigilant about who and/or what is crossing the runways and taxiways.

CASE 4

Avoiding Deer

ASRS accession number: 289965

Month and year: November 1994

Local time of day: 1201 to 1800

Facility: X58, Indiantown Airport (FIG. 6-4)

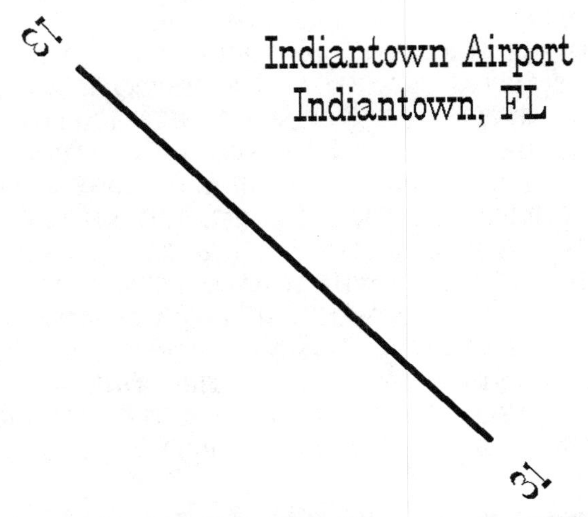

6-4 *Indiantown Airport (this airport diagram is not suitable for navigational purposes).*

Location: Indiantown, FL

Flight conditions: VMC

Aircraft 1: Small GA type single (Cessna 152)

Pilot of aircraft 1: Instructor, 676 hours; student

Reported by: Pilot

Incident description: Runway incursion by deer

Incident consequence: Damage to aircraft

NARRATIVE TAXIED (C-152) ONTO THE DEP END OF RWY 31 AT XX58. WHILE DEMONSTRATING A SOFT FIELD TKO TO MY STUDENT, 2 DEERSIZED ANIMALS RAN ONTO THE RWY ABOUT MIDFIELD. ONE OF THE ANIMALS PROCEEDED SBOUND DOWN THE RWY CTRLINE. AFTER BRAKING THE GND (THE MOMENT THE ANIMALS APPEARED) I ABORTED THE TKO AND DEVIATED R OF THE CTRLINE TO AVOID COLLISION WITH THE APCHING ANIMAL. DUE TO THE RWY CONDITION AND THE ACFT SPD, BRAKING ACTION WAS POOR AND DIRECTIONAL CTL DIFFICULT. THE ACFT ENTERED INTO A SLIGHT HYDROPLANE DUE TO THE MOISTURE IN THE GND. I PULLED THE THROTTLE TO IDLE (AT THE MOMENT THE ANIMALS APPEARED) AND APPLIED SMOOTH AND EVEN BRAKING. TO PROTECT THE NOSEWHEEL, I APPLIED JUST ENOUGH BACK PRESSURE AS NOT TO OBSTRUCT FORWARD VISIBILITY. TIRES BORDERING THE RWY WERE NOT ALIGNED WELL. THE TIRE WERE PAINTED WHITE ON ONE SIDE, BUT SOME OF THEM THAT WERE MISALIGNED WERE BLACK SIDE UP, WHICH MADE THEM DIFFICULT TO SEE. DUE TO THE ANIMALS CAUSING A COLLISION HAZARD, POOR RWY CONDITION AND IMPROPER RWY ORGANIZATION, I STRUCK ONE OF THE TIRES DURING THE MY ABORTED TKO AND DIVERSION. THE TIRE STRUCK AND DAMAGED THE UNDERSIDE (GEARBOX) AND R HORIZ STABILIZER.

SYNOPSIS DURING ABORTED TKO TO AVOID ANIMALS ON RWY, ACFT STRUCK RWY EDGE MARKER TIRE AND WAS DAMAGED.

Postincident Analysis

The pilot and student did not see the deer on the runway until the takeoff was initiated. Preventing impact with the deer required an evasive maneuver resulting in damage to the aircraft.

Several problems are noted from this report.

Problem 1. Lack of visual vigilance by both pilots. The deer did not suddenly materialize on the runway—they had to come from some place and crossed an open area to get to the runway.

Problem 2. Apparent lack of fencing or other controls to keep animals off the runway. Although expensive to install and maintain, this is the only sure means of controlling animal access to the runway.

Problem 3. Tires along the edge of the runway. Many small airports around the country have the runways edged with aircraft tires painted white. Aside from the local FBO trying to make a statement about the number of tires sold over the years, there is no reason for any object being placed along the sides of the runway. The edges should be painted or otherwise marked. Most likely there would have been no damage to the aircraft if the tires had not been there.

Lessons Learned

During the instructional process, there is often considerable talk between the instructor pilot and the student pilot. In this case, there may have been more talk and less attention to the outside of the cockpit than was necessary. The deer should have been visible from several hundred feet at a minimum.

Lesson 1. You cannot depend on there being fencing around airports because it is expensive to install and maintain, even though it would have the advantage of keeping out the animals. Rather, you must be vigilant at all times.

Lesson 2. Tires and other aircraft-damaging objects have no place along the sides of a runway.

CASE 5

Never Saw It

ASRS accession number: 429888

Month and year: March 1999

Local time of day: 1801 to 2400

Facility: PIA, Greater Peoria Regional Airport (FIG. 6-5)

Location: Peoria, IL

Flight conditions: VMC

Aircraft 1: Small GA twin (Piper Aztec)

Pilot of aircraft 1: Single pilot, 2050 hours

Reported by: Pilot

Incident description: Struck object in parking area

Incident consequence: Damage to aircraft

NARRATIVE ON THE MORNING OF MAR/TUE/99, I PARKED A PIPER AZTEC IN A LEGITIMATELY MARKED TIE DOWN SPOT AT THE FBO IN PEORIA, IL. I LAID OVER UNTIL XA55 LCL TIME, LOADED FREIGHT AND PLANNED FOR MY SCHEDULED DEP. AFTER RECEIVING MY IFR CLRNC AND TAXI INSTRUCTIONS FROM PEORIA CLRNC DELIVERY, I STARTED FORWARD IN A L TURN. I FELT THE R PROP STRIKE SOMETHING AND IMMEDIATELY SHUT IT DOWN WITH THE MIXTURE CTL. I THEN SHUT DOWN THE ACFT COMPLETELY AND EXITED THE ACFT TO SEE WHAT HAD HAPPENED AND FOUND THAT THE PROP HAD STRUCK A MOBILE HELIPAD THAT WAS SITTING UNMARKED AND UNLIGHTED IN AN AREA OF THE RAMP MARKED FOR TIE-DOWN. I FEEL THAT A PLT NEEDS TO BE

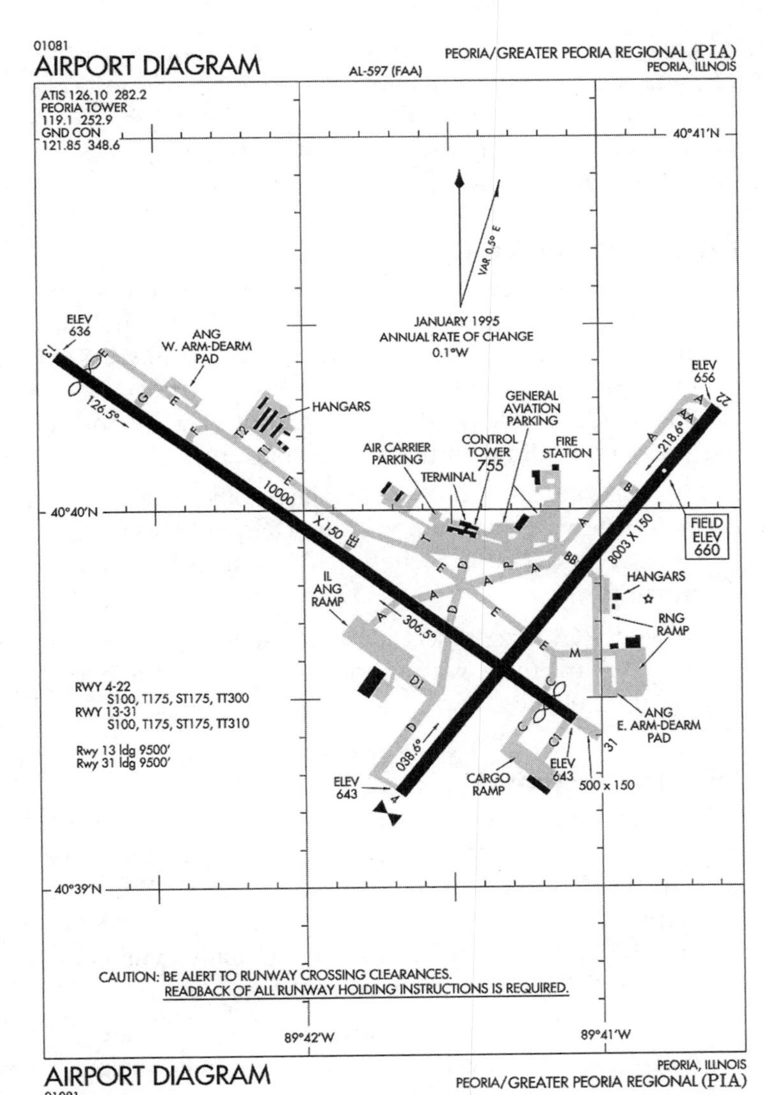

6-5 *Greater Peoria Regional, PIA (this airport diagram is not suitable for navigational purposes).*

AWARE OF HIS/HER SURROUNDINGS, BUT HAVING LOW PROFILE OBJECTS IN POORLY LIGHTED AREAS MARKED FOR SOMETHING ELSE DOES NOT HELP THE SIT.

SYNOPSIS PA23 PLT STRUCK MOBILE HELIPAD ON INITIAL TXWY AT AN FBO IN PIA.

Postincident Analysis

Obviously, the pilot never checked the aircraft's surroundings. Had the area been checked, the helipad would have been noticed, and there would have been no incident, unmarked and poor lighting or not.

This incident falls on the pilot for not visually checking the surroundings in the aircraft parking area:

Problem 1. Failure to observe a large object in the immediate vicinity of the aircraft. The pilot should have seen the helipad when he first approached the airplane.

Problem 2. A large object left in an aircraft parking area. Perhaps there was a more appropriate place to park the helipad.

Lessons Learned

A pilot must be aware of his or her aircraft's surroundings. Just observing during a walk-around would have shown the pilot that the helipad was nearby. You cannot depend upon others to do what is always correct or proper. Therefore, you must be visually vigilant—whether in the airplane or around the airplane.

Lesson 1. Watch where you are and where you are going—even if not on a runway. Be vigilant, vigilant, vigilant!

Lesson 2. Check your surroundings before moving the aircraft. Be vigilant, vigilant, vigilant!

CASE 6

I'm Busy Working

ASRS accession number: 425117

Month and year: January 1999

Local time of day: 1201 to 1800

Facility: AXH, Houston Southwest Airport (FIG. 6-6)

Location: Houston, TX

Flight conditions: VMC

Aircraft 1: Several and various

Pilot of aircraft 1: Several and various

Reported by: Pilot (not flying at the time)

Incident description: Runway incursion by survey crew

Incident consequence: See narrative

NARRATIVE AT APPROX AB00 LCL TIME I WAS WITNESS TO AN EVENT (OR MAY IT BE CALLED A SERIES OF EVENTS) WHICH WAS DEFINITELY A SERIOUS SAFETY ISSUE. I WAS PREFLTING FOR A LCL VFR FLT WHEN I NOTICED A CAR DRIVING UP AND DOWN THE RWY (RWY 9/27) AT HOUSTON SOUTHWEST ARPT (AXH). THERE WAS QUITE A LOT OF LCL TFC BECAUSE THE WX WAS BEAUTIFUL. THE CAR STOPPED ON A TXWY, WHICH IS THE ONLY ENTRY/EXIT POINT FOR A SET OF T HANGARS, AND A GENTLEMAN GOT OUT, OPENED THE TRUNK AND

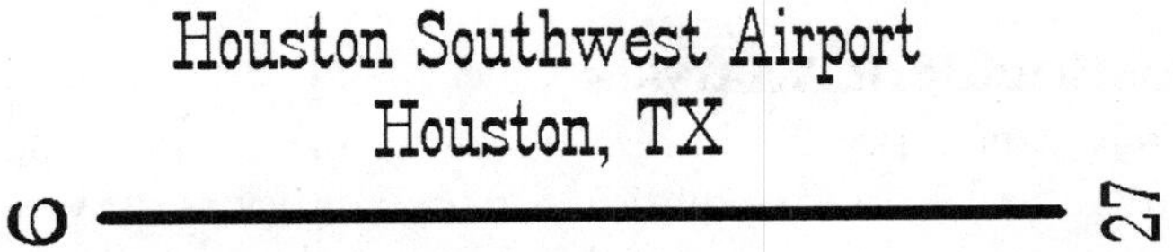

Houston Southwest Airport
Houston, TX

9 27

6-6 *Houston Southwest Airport (this airport diagram is not suitable for navigational purposes).*

EXTRACTED A SURVEYOR'S TRANSIT/TRIPOD. HE WALKED OUT TO THE CTRLINE OF THE RWY WITH HIS EQUIP AND SET IT UP ON THE RWY AT THE APCH END NEAR THE RWY END LIGHTS. I APCHED THE MAN AND ASKED ABOUT THE OP, WHETHER IT WAS NOTAMED, IF THE ARPT WAS CLOSED, ETC. I WAS TOLD THAT HE WAS 'WORKING,' THAT THERE WAS NO NOTAM BECAUSE LCL PLTS NEVER CHKED THEM, THAT HE HAD 50 YRS EXPERIENCE AND NOT TO BOTHER HIM. I VOICED MY CONCERNS ABOUT HIS SAFETY, THE SAFETY OF THOSE WORKING WITH HIM AND THE SAFETY OF THE PLTS, ACFT AND PAX OPERATING AT THE ARPT THAT DAY. HE STATED THAT HE HAD 'WORK' TO DO AND TO LEAVE HIM ALONE. I WITNESSED SEVERAL OTHER PLTS VOICE THEIR SAFETY CONCERNS AS WELL. ALSO, I WITNESSED GARS, LONG LNDGS AND OTHER ABNORMAL APCHS/LNDGS. WHEN I LEFT THE ARPT AT APPROX AD30 LCL TIME THE SURVEYING CREW WAS STILL ON THE RWY (WITH INDIVIDUALS WALKING UP AND DOWN THE CTRLINE BTWN ACFT OPS) AND THEIR VEHICLES WERE PARKED ON A RWY EXIT TXWY COMPLETELY BLOCKING THE EXIT AND WITH 1 VEHICLE EXTENDING ONTO THE RWY. CORRECTIVE ACTIONS: ARPT MGR TO HAVE EXERCISED CTL OVER THIS OP—FROM FILING THE NOTAM TO PERSONALLY OVERSEEING THIS ENTIRE SIT.

SYNOPSIS A PLT RPT ON THE UNCOORD USE OF GND EQUIP AT AXH, TX. A SURVEYOR'S CREW SET UP SHOP ON THE ACTIVE RWY WITH TFC HAVING TO BE CREATIVE SO AS TO MISS THE MISCELLANEOUS EQUIP ON THE RWY. ARPT MGR OF NO HELP.

Postincident Analysis

What was reported here was very serious because it affected all airport operation for several hours. It was an activity of arrogance on the part of the surveyor, as shown in the report by the statement, "I WAS TOLD THAT HE WAS 'WORKING,' THAT THERE WAS NO NOTAM

BECAUSE LCL PLTS NEVER CHKED THEM, THAT HE HAD 50 YRS EXPERIENCE AND NOT TO BOTHER HIM."

The legal ramifications, had an individual been struck on the ground or an aircraft damaged due to landings to accommodate the crew working on the runway, could have been far-reaching. Yet the surveyor further stated, "THAT HE HAD 'WORK' TO DO AND TO LEAVE HIM ALONE."

Several problems are noted from this report:

Problem 1. Based on the statement, "THEIR VEHICLES WERE PARKED ON A RWY EXIT TXWY COMPLETELY BLOCKING THE EXIT AND WITH ONE VEHICLE EXTENDING ONTO THE RWY," the airport should have been closed while this activity was taking place.

Problem 2. The airport management was negligent in not taking action when the problem was brought to their attention. The synopsis of the report says, "ARPT MGR OF NO HELP."

Problem 3. Pilots continued to fly from and to the airport, regardless of what they saw on the ground, "HAVING TO BE CREATIVE SO AS TO MISS THE MISCELLANEOUS EQUIP ON THE RWY."

Lessons Learned

You cannot count on airport management to do what is correct. However, when there is an obstacle to flying, whether to takeoff or to landing, the prudent pilot would avoid the problem. In this case, this would mean not using the runway until the problem abated.

Several lessons can be learned from this report:

Lesson 1. Do not count on others to do what is right or required.

Lesson 2. Grumbling and getting into arguments with work crews (the surveyor in this case) is unlikely to produce the desired results. Call the local Federal Aviation Administration (FAA) office—immediately!

Lesson 3. Continued flying under these conditions is accepting the additional risks being placed on you. Remember that the commander of an aircraft is responsible for everything. Just because there was no NOTAM about the airport would not release a pilot from blame if there had been a runway incident.

CASE 7

Road rage

ASRS accession number: 421476

Month and year: November 1998

Local time of day: 1201 to 1800

Facility: MEM, Memphis International Airport (FIG. 6-7)

Location: Memphis, TN

Flight conditions: VMC

Aircraft 1: SF-340

Aircraft 2: DC-9

Pilot of aircraft 1: Captain, 9500 hours; first officer

Pilot of aircraft 2: Captain

Reported by: Pilot of aircraft 1

Incident description: Taxiway incursion

Incident consequence: Verbal altercation

NARRATIVE OUR ACFT LANDED AND TAXIED CLR OF RWY 27 AND RECEIVED A TAXI CLRNC TO GATE. WE PROCEEDED S ON TXWY N TO APPROX ABEAM GATE, AT WHICH TIME GND TOLD US TO GIVE WAY TO 2 OPPOSITE DIRECTION SAABS GOING IN TO GATE. WE COMPLIED BY SLOWING OUR TAXI AND ONCE CLR, WE RESUMED OUR NORMAL TAXI, S ON TXWY N. ABEAM GATE XY, I SAW A GND VEHICLE IN THE SW CORNER OF TXWYS M AND M5

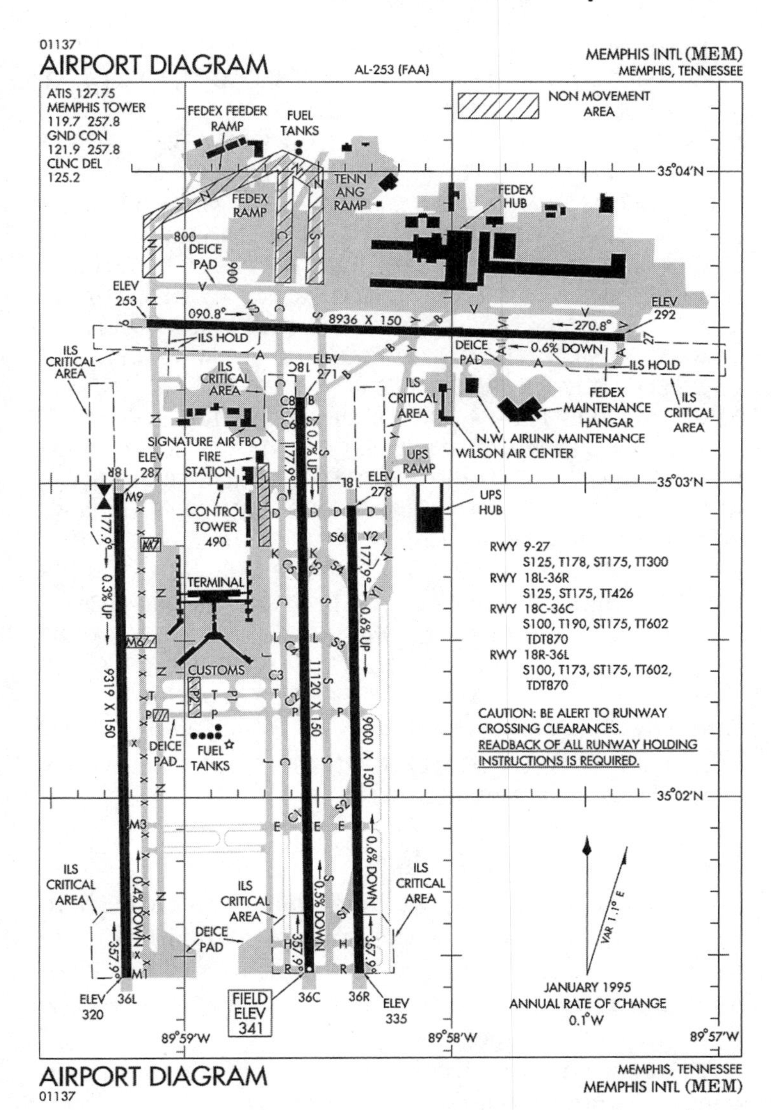

6-7 *Memphis International, MEM (this airport diagram is not suitable for navigational purposes).*

AND AT ABOUT THE SAME TIME HEARD HIM REQUEST PERMISSION TO GO OUT ON RWY 36L AFTER FOD. AT ALMOST THE SAME TIME A DC9 CLRED RWY 36L AT TXWY M5 AND REQUESTED TAXI CLRNC TO TXWY B29, WHICH HE RECEIVED PROMPTLY FROM GND CTL. (THE GND CTLR THEN BEGAN CONVERSING WITH THE GND VEHICLE ABOUT HIS INTENTIONS.) THE DC9 NEVER SLOWED DOWN AFTER TURNING OFF RWY 36L AND RECEIVING HIS TAXI CLRNC. HE CAME RIGHT ACROSS TXWY M AND WAS APPARENTLY GOING TO CROSS TXWY N AS WELL. IT WAS AT THIS POINT THAT I BEGAN TO SLOW BECAUSE IT LOOKED LIKE HE WAS GOING TO PULL OUT ACROSS TXWY N DIRECTLY IN FRONT OF US. AT THE LAST SECOND HE SLOWED DOWN TO STOP AND WE CONTINUED IN TO THE GATE WITHOUT FURTHER INCIDENT. SHORTLY AFTER THE PAX DEPLANED, THE DC9 CAPT APPEARED AT OUR ACFT AND TRIED TO ENGAGE ME IN A SHOUTING MATCH (HE WAS FULLY INTO 'ROAD RAGE' AT THIS POINT). I DECLINED AND TOLD HIM TO EXIT OUR ACFT. I BELIEVE THIS WHOLE SIT WAS CAUSED BY THE GND CTLR'S DISTR INVOLVING THE GND VEHICLE AND THE DC9 CAPT'S FAILURE TO LOOK BOTH WAYS IN A TIMELY FASHION BEFORE XING A MAJOR (BUSY) TXWY.

SYNOPSIS TXWY DISPUTE TURNS TO CONFRONTATION BTWN SF340 AND DC9 FLCS AT MEM ARPT.

Postincident Analysis

Well, here it is at the airport. Road rage has struck! There just isn't the time to slow down for a moment. No time for a little patience. Just have to get there first. Plain tough for those in the way. That's the attitude of today: "All right now for me!"

This report is very disturbing because it shows an incident at an airport that has strong parallels with what happens on overcrowded highways. It isn't tolerated on the highways, and it shouldn't be tolerated at an airport.

Simply stated, road rage is caused by a sense of crowding, pushing and/or being pushed, and the perception of a need to rush and is a demonstration of lack of self-control.

Problems noted from this report include what may have led up to this road rage type of incident:

Problem 1. The airport was very busy—hence a sense of crowding and a perceived need to rush.

Problem 2. The controller used valuable radio air time for conversing with the ground vehicle, adding to perception of a need to rush..

Problem 3. The pilot of the DC-9 may have felt that being in the larger aircraft allowed the intimidation of other aircraft—pushing and/or being pushed.

Problem 4. Lack of self-control on the part of the DC-9 pilot. The verbal confrontation with the other flight crew.

Lessons Learned

There is no place in aviation for a pilot with road rage. Verbal confrontation is only a single short step from physical altercation—which could have been demonstrated by an assault on the SF-340 pilot or by deliberately ramming the DC-9 into the smaller aircraft. Either action was not out of the realm of possibilities.

In life, you do the best you can to avoid situations and people like this angry pilot, but you may not always be successful. The pilot of the SF-340 did what was correct in ordering the upset pilot from the aircraft. This probably avoided further escalation of an already bad situation.

At that point it probably would not have been a bad idea to contact the company for guidance about reporting the offending pilot to the FAA, since there is little doubt that this type of anger has appeared prior to this incident and will appear again—perhaps with far more serious results.

Lesson 1. Heads up! The SF-340 flight crew was watching and thus was able to stop, had the DC-9 continued across the taxiway. It is prudent to wait for another aircraft if that will prevent you from being involved in a collision.

Lesson 2. Relax! The flight crew has very little control over the surrounding traffic. Go with it—it sure is easier than fighting it.

Lessons Learned in this Chapter

This chapter has closely examined a number of unusual runway and taxiway incursions and incidents—as reported on the ASRS reports used for the case studies. These are case studies that should make you stop and think that if anything can go wrong, it will.

In each case study several lessons were learned. Of these lessons, the following appears to be the most common:

Visual vigilance. Keep heads out of cockpits. Look around. Watch for other aircraft activity on the airport. Watch for any other activity on the airport, such as animals, vehicles, etc. As seen, it can be that "other" activity that will get you. Everything and everybody cannot be fenced out or avoided or lighted. Look out for yourself.

To look around and listen around cannot be emphasized enough—vigilance to all!

Part 3
Prevention of Runway Incursions

Up to this point, this book has been a reporter of past events. Runway incursions have been described and explained. How airports operate has been described. And we have looked at the players in the cockpit and in the tower. Along the way, many Aviation Safety Reporting System (ASRS) reports were read and commentary made. Now it's time to look to the future and learn how runway incursions can be prevented.

7

Preventing Runway Incursions for Pilots

As we said in Chapter 1, "Runway incursions are not caused just by pilots, or air traffic controllers, or vehicle operators, or pedestrians. Runway incursions are a problem that all of us in the aviation community share—and must solve."

A Cooperative View

Pilots are encouraged to visit air traffic facilities (towers, centers, and flight service stations, or FSSs) and familiarize themselves with the air traffic control (ATC) system. Note that on rare occasions some facilities may not be able to approve a visit because of ATC workload or other reasons. Thus it is suggested that pilots contact the facility of interest prior to a planned visit.

Visiting groups are more efficient for personnel of ATC to work with because the information will reach more people at the same time. When contacting ATC for a visit, indicate the number of persons in the group, the time and date of the desired visit, and the primary interest (types of pilots) of the group. This information allows the facility to prepare an informative and worthwhile visit.

Visiting the control tower of a busy airport can be a very enlightening lesson for a pilot. Seeing air traffic through the controllers' eyes lends a completely different perspective to the movement of aircraft on the airport. Such visits are highly recommended.

Recommended Areas of Improvement

Now it is time to look at methods of preventing runway incursions. The following recommendations and suggestions come from many sources and are directed toward pilots—pilots of general aviation aircraft, military aircraft, and commercial transports.

The three areas for recommended improvement by pilots are (first seen in Chapter 1):

1. Improved communications
2. Increasing airport knowledge
3. Proper cockpit procedures

In the following pages, each of these areas for recommended improvement are expanded on.

Improved Communications

Check that you have your radio(s) tuned to proper frequency, volume, and squelch settings. The time to prepare your radios for use is before they are needed—not when the need arises.

Keep your radio communications with ATC clear and concise. Speak slowly and clearly, using common phraseology (see Appendix A for the Pilot/Controller

Glossary). It is imperative that queries and instructions are understood. There is no room for any confusion when receiving instructions from ATC, making a read-back, in acknowledgments, or querying ATC.

Listen before you transmit to avoid doubling with another airplane. *Doubling* occurs when two pilots are talking on the same frequency at the same time, which causes radio transmissions to become garbled—leading to confusion caused by missed queries and instructions. *Blocked radio transmissions* is the term for the garbled transmissions resulting from doubling.

Only talk to ATC when necessary. The controller does not have time for a friendly chat, and other pilots will not be impressed by your friendliness. Unnecessary radio usage takes away from others. Time and frequency are finite, having limits placed on them. Use the radio efficiently—as needed for safe flight operations.

Monitor all the radio traffic at the airport so that you can develop a mind picture of all airplane activity. Listening to other requests and clearances will tell you who else is using the airport: who is about to land and on what runway, what traffic to expect along the taxiway, and who is moving toward the run-up area.

Ensure that you understand all your instructions from ATC. Do not assume that you got it right the first time you heard it. If there is any doubt, ask the controller for a repeat. "Say again" is the phrase you should use when requesting a repeat of a clearance of instruction.

Read back all runway/taxiway hold-short instructions verbatim—allowing for a double check. This requirement is printed on some airport diagrams. Know what to do if you are at a controlled airport and your radio system fails (see Chapter 2).

From the *Aeronautical Information Manual* (AIM)

The AIM, Section 4, has a few paragraphs concerning the very basics of radio communications. Although general in nature, the recommendations stand for all pilots and are quoted here:

4-2-2. Radio Technique

(a) **Listen** before you transmit. Many times you can get the information you want through ATIS or by monitoring the frequency. Except for a few situations where some frequency overlap occurs, if you hear someone else talking, the keying of your transmitter will be futile and you will probably jam their receivers, causing them to repeat their call. If you have just changed frequencies, pause, listen, and make sure the frequency is clear.

(b) **Think** before keying your transmitter. Know what you want to say and if it is lengthy; e.g., a flight plan or IFR position report, jot it down.

(c) The microphone should be very close to your lips and after pressing the mike button, a slight pause may be necessary to be sure the first word is transmitted. Speak in a normal, conversational tone.

(d) When you release the button, wait a few seconds before calling again. The controller or FSS specialist may be jotting down your number, looking for your flight plan, transmitting on a different frequency, or selecting the transmitter to your frequency.

(e) Be alert to the sounds *or the lack of sounds* in your receiver. Check your volume, recheck your frequency, and *make sure that your microphone is not stuck* in the transmit position. Frequency blockage can, and has, occurred for extended periods of time due to unintentional transmitter operation. This type of interference is commonly referred to as a "stuck mike," and controllers may refer to it in this manner when attempting to assign an alternate frequency. If the assigned frequency is completely blocked by this type of interference, use the procedures described for en route IFR radio frequency outage to establish or reestablish communications with ATC.

(f) Be sure that you are within the performance range of your radio equipment and the ground station equipment. Remote radio sites do not always transmit and receive on all of a facility's available frequencies, particularly with regard to VOR sites where you can hear but not reach a ground station's receiver. Remember that higher altitudes increases the range of VHF "line of sight" communications.

Radio Phrases

There are times when you need to make requests to ATC for clarification purposes. The following phrases, also found in the AIM Pilot/Controller Glossary, may be appropriate during specific conditions to eliminate clearance confusion:

Speak slower. This is used in verbal communications as a request to reduce the speech rate.

Read back. This is used as a request for the controller to repeat your message back to you.

Say again. This is used to request a repeat of the last transmission or part thereof.

Words twice. This is used as a request meaning that communication is difficult and for the controller to say every phrase twice.

Note that the controller is likely to be less than pleased with your request, but do not let this deter you from asking for help. Slowing down communications by asking for any of the preceding will upset some controllers because they feel a rush to keep the air traffic moving. However, they are required to comply with your request. Unfortunately, you may feel the warmth from the controller in your headphones as the clearance comes slower.

In the same vein, it has been rumored that a senior airline pilot, after receiving a long clearance from a fast-talking controller, added fuel to the fire by saying, "Copy cleared to FL 230." No doubt there was a repeat of the initial clearance—probably deliberately slow and dripping with sarcasm—but the incident probably garnered the controller's undivided attention.

Increased Airport Knowledge

You must know where you are, where you are going, and how to get there. This applies to driving, yachting, hiking, and flying. And maps are used for each of these activities—to prevent you from becoming lost or from going someplace you did not intent to go.

Most controlled airports have numerous taxiways, runways, and intersections. It can be a daunting task to "get there from here" unless you know your way around the airport. Knowing your way around an airport is why you should use airport diagrams. They are your maps of the airport.

Detailed airport diagrams show the ground plan, including runways, taxiways, buildings, and other important information required for safe ground movement. Although some pilots think of airport diagrams as useful only for IFR operations, they are extremely useful for VFR operations also.

Usage Warning

The National Aviation Charting Office (NACO) warns that the use of obsolete charts or publications for navigation may be dangerous. Always be sure that the charts and diagrams you are using are current. Also consult Notices to Airmen (NOTAMs) for changes to publications that may occur during the effective dates of publication.

Free Airport Diagrams

Individual airport diagrams, in Adobe PDF format for printing, are available online, without cost, at *http://acc.nos.noaa.gov/AirportDiagrams.html.* Adobe-formatted documents require a special reader for use. That reader, Acrobat Reader, is available for free download from the site just mentioned and from many other sites on the Internet.

The simplicity of downloading only those airport diagrams you specifically need cannot be beaten. It is timely, cost-effective, and easy. Depending on the printer used and its setup, the clarity of the downloaded diagrams can exceed that of the purchased diagrams. A laser printer is recommended for printing airport diagrams because the printed output is more lasting (waterproof) than inkjet output.

Purchased Airport Diagrams

Airport diagrams are available for purchase from NACO as the *Airport/Facility Directory* (published softbound in seven volumes by geographic region) at

FAA Distribution Division
AVN-530
National Aeronautical Charting Office
Riverdale, MD 20737-1199
Phone: (301) 436-8301, or (800) 638-8972 toll free, U.S. only

FAX: (301) 436-6829
Online: *http://acc.nos.noaa.gov/Catalog/afd.html*

As a side note, on October 1, 2000, the Federal Aviation Administration (FAA) created the National Aviation Charting Office (NACO) as a part of Aviation System Standards (AVN). This is the result of an agency transfer of the Office of Aeronautical Charting and Cartography (AC&C) from the National Oceanic and Atmospheric Administration (NOAA). Users of NACO products began seeing the Department of Transportation and FAA logos on the former NOAA charts beginning with the October 5, 2000, charting effective date.

There are also commercial vendors that sell products that are similar to those offered by NACO. Both NACO and commercial vendor products are available for purchase at most FBOs and from various online sources, including Sporty's Pilot Shop at *http://www.sportys.com/shoppilot/*.

Using the Airport Diagrams

Before departure or landing, review the appropriate airport diagrams, and then keep them handy for immediate reference. This will allow you to be prepared for your taxi instructions and give you a visual aid while taxiing. Always check the taxi route the controller gives you with the diagram. If you have questions, concerns, or doubts, ask the controller.

Proper Cockpit Procedures

The flight crew's full time and attention are required during flight operations to prevent runway incursions. A proven effective cockpit procedure providing for full time and attention is the *sterile cockpit*. Maintaining a sterile cockpit environment means avoiding unnecessary conversation during ground, takeoff, and landing operations—allow nothing to interrupt your concentration with the task at hand.

Be aware of your surroundings by constantly looking around for other traffic—airplanes, vehicular, pedestrian—and be sure your airplane is visible to others (properly lighted). Know where you are supposed to be, and be there.

What If

What if something goes wrong? For example, you are taxiing your airplane from a parking area to an active runway or from a runway to a parking area—and you become lost. Have a plan—a very simple plan that will work in all circumstances!

For example, if you become disoriented (lost), stop and contact ATC immediately. Ask for progressive taxi instructions, if needed. ATC will talk you through the maze of getting there from here. Although you may feel that you are a burden to the controllers, you will be more of a burden if there is an incident. If you need help, get help. Do not merely blunder on. The situation most likely will deteriorate without help.

Note: Never stop on an active runway after landing to ask for directions. Clear the runway—beyond the hold lines—then stop and ask ATC for directions.

Paying Attention

When we are tired, whether from overwork or from long hours of activity, and as the workload and distractions increase, attention to communications tends to decrease. In the long term, fatigue levels increase; in the short term, mistakes happen. If you are tired, get some rest.

Boredom is just as detrimental to paying attention as is fatigue. Closely related to boredom is complacency. Both factors can allow a lack of attention to provide plenty of space for error.

Do not let yourself get to the point of having done it so often that you can do it with your eyes closed because just when you are about to do it as you always have, someone will throw a curve—and you may not catch it. You can easily become a victim of hearing what you are expecting and missing what you were really told.

Reduced Visibility

As visibility decreases, it becomes increasingly difficult for pilots and controllers to make/maintain visual contact. When experiencing reduced visibility, pilots should realize that cockpit workload and distractions tend to increase. The more workload and distractions, the more chance there is for error.

Nonstandard Procedures Requested by ATC

Controllers at busy airports sometimes use a variety of techniques to keep traffic flowing smoothly. As a result, you may be asked to do any of the following:

Fly a faster than normal final approach. To get on the ground and out of the way more quickly.

Extend your downwind. To delay your landing.

Switch to another runway at the last minute. To accommodate other traffic.

Do a 360-degree turn or S-turns on final. To allow traffic ahead to clear.

Taxi around another aircraft in the run-up area. To expedite your takeoff.

Be proficient with your airplane so that you can handle such unusual procedures safely. However, remember that you are the pilot in command and have the right to decline any procedure that may put your safety in jeopardy.

You should cooperate with ATC as much as possible, but do not be afraid to refuse an unusual request initially or to change your mind if the procedure starts to fall apart. The pilot is the aircraft commander and does have the final say—and responsibility.

How the Pros Do It

There is great respect for professional pilots and their methods of operation. They have extensive training and experience. Let's take a brief look at how their standard procedures aid in runway incursion prevention.

Most airlines and corporate flying departments have detailed procedures for flight operations at specific controlled (towered) airports. These procedures are far more detailed than the simple requirements of the Federal Aviation Regulations (FARs) most general aviation pilots are required to adhere to.

The following lists of procedures are not complete or exhaustive, but they serve to illustrate some useful techniques that will make ground operations more professional and safe.

For Takeoff

1. The airport diagram is reviewed before starting the engines, and the aircraft's position relative to the active runways is noted.
2. The captain will handle the aircraft initially.
3. The first officer will contact ground control prior to aircraft movement for taxi instructions. The first officer will copy the taxi instructions and reference the airport diagram—asking for repeats as necessary.

4. During all communications with ATC, the flight crew will listen to what is being said (while other cockpit duties wait). This step is to ensure that there is agreement among the flight crew as to what they heard.
5. A full read-back of all ATC instructions is required.
6. A taxi chart (airport diagram) is used while taxiing.
7. Both pilots will listen to ATC and visualize the positions of other aircraft.
8. Other cockpit duties stop when crossing a runway. Both pilots must agree that ATC has cleared them to cross the runway, and both pilots must check the runway visually in both directions prior to crossing.
9. Taxi lights will be used day and night as a warning to other aircraft.
10. Landing and strobe lights are turned on at the beginning of the takeoff roll.

For Landing

1. The flight crew will listen to ATIS information and review the airport diagram.
2. The first officer will contact ATC, inform ATC of ATIS reception, and ask for landing instructions.
3. The landing clearance will be copied and referenced to the airport diagram. Repeats will be requested from ATC as needed.
4. During all communications with ATC, the flight crew will listen to what is being said (while other cockpit duties wait). This step is to ensure that there is agreement among the flight crew as to what they heard.

5. A full read-back of all ATC instructions is required.
6. Both pilots will listen to ATC and visualize the positions of other aircraft.
7. Landing and strobe lights are turned on at 10 miles out.
8. After landing, a taxi chart (airport diagram) is used while taxiing.
9. Other cockpit duties stop when crossing a runway. Both pilots must agree that ATC has cleared them to cross the runway, and both pilots must check the runway visually in both directions prior to crossing.
10. Taxi lights will be used day and night as an warning to other aircraft.

An explanation and many recommendations have been made toward preventing runway incursions. However, what should be done when an incursion does happen? Next, reporting a runway incursion via an Aviation Safety Reporting System (ASRS) report is explained.

Making ASRS Reports

If you are involved in a runway incursion or near incursion, report the incident via the NASA Aviation Safety Reporting System (ASRS). ASRS uses the information on the forms to develop a database. You will not receive an FAA violation if you report an incident to ASRS.

Report Specific Airport Problems

Confusing or deteriorating runway/taxiway surface markings; faded, damaged, or otherwise unreadable signs; and inaccurate airport diagrams all can lead to runway incursions. At any time you encounter confusing, deteriorating, unreadable, or inaccurate surface navigation aids, you should report them to ATC. In addition, a report to the ASRS is strongly recommended. ASRS maintains a database of reported hazards.

Alert messages, based on pilot reports made to ASRS, are forwarded to the appropriate airport authorities for action. Airport authorities are requested to provide responses to ASRS reports. These responses serve as an important check on the types of corrective actions being taken and closes the loop in the incident-reporting process.

ASRS report forms are available from NASA as follows:

ASRS
635 Ellis St., Suite 305
Mountain View, CA 94043
FAX: (415) 967-4170
Online: *http://asrs.arc.nasa.gov/forms_nf.htm/*

Mail your completed form(s) to:

NASA Aviation Safety Reporting System
Post Office Box 189
Moffett Field, CA 94035-0189

At this time, there are no provisions in the program for FAX or e-mail submission of ASRS reports—although this feature may be coming soon.

Remember that as a reporting person, you are protected by Federal Aviation Regulations (FARs), Section 91.25, the Aviation Safety Reporting Program, which prohibits any use of the ASRS for enforcement purposes. The FARs state, "The Administrator of the FAA will not use reports submitted to the National Aeronautics and Space Administration under the Aviation Safety Reporting Program (or information derived therefrom) in any enforcement action except information concerning accidents or criminal offenses which are wholly excluded from the Program."

Flight Review Requirement

If you are already a rated pilot, you have not escaped the FAA's new agenda. Flight instructors have been told to verify that every pilot, during a flight review or any type of training, can demonstrate an understanding of airport signage, lighting, and taxiway/runway surface markings.

Let Us Review

So far in this book we have looked at facts, figures, scenarios, rules, and examples regarding runway incursions. Relative to preventing these incursions, suggestions and recommendations include

Understand airport signage, markings, and lighting.

Turn on aircraft lights (including rotating beacons and/or strobe lights) while taxiing.

Do not hesitate to request clarifying or progressive traffic instructions from ATC when you are unsure of the taxi route assigned to you.

Always check for traffic (airplane, vehicular, pedestrian, etc.) before crossing any runway or entering or crossing a taxiway—even after receiving an ATC clearance.

Always clear the active runway as quickly possible and wait for taxi instructions (if not already received) before any further movement.

Keep in mind that when things get busy in the cockpit, stay focused on flying the airplane. Do not try to do too many things at once, such as shuffle through charts, read the airport diagram, tune the radios, program the GPS, configure the aircraft, and copy a clearance. A single pilot can be quickly overwhelmed with the workload, possibly resulting in missed critical information and a disastrous runway incursion.

Review the appropriate Notices to Airmen (NOTAMs) for information on runway and/or taxiway closures and construction at destinations and along the route.

Regarding land and hold-short operations (LAHSO), remember that the pilot in command of the airplane has the final authority to accept or decline any land and hold-short clearance. You should decline a LAHSO clearance if you are not comfortable with it.

8

FAA Future Plans

Where is the blame to be placed? Or does blame have to be placed at all? Perhaps the best approach is for the major players to work on solutions—toward the goal of reducing or eliminating runway incursions.

To this end, in June 2001, Aircraft Pilots and Owners Association (AOPA) President Phil Boyer told a congressional committee that simple solutions and pilot education are the keys to reducing the number of runway incursions. He stated that although the numbers of runway incursions for general aviation (GA) appeared disproportionately high, they actually indicated mostly minor events—with little or no chance for collision. Mr. Boyer also stated that "technology is not the total solution to this problem." He indicated that properly painted runway and taxiway surface markings, improved signage, and pilot education would help greatly in reducing the current number of runway incursions.

At the same time that the president of the AOPA was speaking on specific points about reducing runway incursions, the Air Transport Association (ATA) was highlighting the available technology solutions to the problem. As part of its push for technology solutions, the ATA said, "runway incursions are an urgent safety issue which should be rapidly addressed, in part, through the application of modern technology." The statement went on to reflect the need for more prominent signage and lighting, as well as improved awareness and training for airport personnel.

During the same hearings, the National Air Traffic Controllers Association (NATCA) said, "Today's airport environment has become increasingly complex for pilots and air traffic controllers," continuing that over the most recent 5 years, air traffic has increased in the United States by 27 percent. Stressed was the fact that training is a big part of the overall picture, including work on pilot/controller phraseology.

The Federal Aviation Administration (FAA) is, and has been for several years, exploring various methods of reducing runway incursions. Some methods involve the use of high technology, whereas others use education and enforcement. None to date have been truly successful at curbing the rising number of runway incursions. However, in defense of this lack of success, it is only fair to remind readers that the amount of air traffic at the busier airports is continuing to grow—thereby providing the opportunity for an increasing number of incursions.

Technology Programs

The FAA is involved in the development and deployment of several high-tech systems designed to reduce runway incursions. Not all programs (systems) are installed everywhere. In fact, the majority of the systems are installed at only a few locations for system evaluation purposes. No particular system has yet been chosen as the final product for use in reducing runway incursions.

Airport Surface Detection Equipment (ASDE-3). ASDE-3 is essentially a ground-based radar system that detects surface movement, calculates a probable intended path, and places that information onto the air traffic controller's radar screen. This radar monitoring of airport surface operations is to aid in the orderly movement of aircraft and ground vehicles—especially during periods of low visibility such as rain, fog, and night operations. The ASDE-3 system, manufactured by Northrop Grumman Corporation, has been installed at 34 of the busiest U.S. airports.

Airport Surface Detection Equipment (ASDE-X). ASDE-X is a lower-cost system that the FAA plans to install at the next busiest 25 airports starting in 2003. The system is intended to give controllers a high-resolution overview of airport operations. It does not, however, have any predictive capabilities The system is not slated for full operation until 2007.

Airport Movement Area Safety System (AMASS). AMASS is an add-on enhancement to the ASDE-3 radar system and provides automated alerts and warnings (visually and aurally) of potential runway incursions and other hazards. AMASS extends the capability of the ASDE-3 system and enhances surface movement safety.

AMASS concentrates on airborne aircraft, 200 ft or less above the ground and within 6000 ft of the runway threshold, and all on-airport surface movements. AMASS hardware is reportedly installed at 33 of the 34 busiest U.S. airports and will go into service at a later time (perhaps as late as 2003).

Critics of AMASS argue that relying on human controllers for action is too passive. Further, they argue that the system is 7 years behind schedule, at almost triple its original estimated cost, and will not do all of what it was originally claimed to do. In fact, Carol J. Carmody, acting chairman of the National Transportation Safety Board (NTSB), testified that AMASS, as currently designed, is inadequate, saying, "AMASS does not appear to be able to provide sufficient warning time to prevent some runway collisions and does not provide direct warnings to flight crews and other vehicle operators."

Runway Incursion Prevention System (RIPS). RIPS is a radar system, more active than the FAA's improved ASDE-3, to be part of a system that NASA is developing to give pilots a clear electronic picture of what's outside their window—a moving map-type display on the control panel in possible combination with a heads-up display (HUD, transparent on the windshield) that will graphically illustrate runways and taxiways and warning of conflicts.

Multilateration systems. Multilateration systems are designed to track all aircraft equipped with mode A/C, mode S (including TCAS), ADS-B, and military IFF transponders at selectable ranges out to more than 200 nautical miles from the airport.

Controllers see the results on a dedicated display screen, with each aircraft tagged with its identification number, altitude, ground speed, and intent (climbing, descending, or level), plus a variety of other selectable data, against a background depiction of the local airspace environment, including airways, radio aids, reporting points, and runway approach paths. At a much larger scale setting, the system's airport surface presentation shows the terminal area, gates, ramps, taxiways, and runways.

The system plots aircraft positions, on the airport surface as well as in the air, with pinpoint accuracy that radar cannot equal. It allows controllers to instantly pick out individual aircraft parked on the ramp, moving along taxiways, lining up at the runway or taking off and landing. The system is completely independent of the local air traffic control (ATC) radars.

Loop Technology (LOT). LOT represents a congressionally initiated development initiative. This initiative requests the FAA to look into the application of inductive loops in an airport environment to reduce runway incursions and as a possible lower-cost alternative to an ASDE at appropriate airports or where there is a localized problem not requiring the full capabilities of an ASDE. An earlier (1993) FAA-sponsored broad agency announcement (BAA) proved the technical feasibility of inductive loops to detect aircraft in an airport environment. That experiment also proved that aircraft passing over an inductive loop produced a unique signature signal that is repeatable and can be used to classify the aircraft (i.e., B737, MD-80, DC-9, Cessna 152, etc.). Additionally, through signal processing and the application of neural-network technology, other information can be extracted from this signature that can be used to perform a tracking function. The objective of LOT is to provide a prototype system to assess the potential operational suitability and performance of distributed loop-based systems to improve surface safety, as well as a tool to aid the controllers' situational awareness.

Runway Status Lights (RWSLs). RWSLs represent a runway incursion development initiative that initially underwent an evaluation at Boston's Logan International Airport. The system is radar-based and is intended to improve on-airfield flight crew situational awareness by providing pilots with a visual advisory of runway status. The system consists of two sets of lights: runway entrance lights (RELs) and takeoff hold lights (THLs), designed to prevent runway incursions. RELs are located at the entrances to runways, and THLs are placed on the runway at the position for the start of the takeoff roll. The RWSLs function as an independent backup (no hands-on operation required) and do not add to ATC workload.

The technology programs directed toward the prevention of runway incursions have been explored. Next, the educational programs directed toward controllers and pilots are discussed.

Prevention Education

Phil Boyer of the AOPA said, "Education, rather than expensive technology, is the best means to prevent runway incursions for GA." He also indicated, when testifying to Congress, that properly painted runway and taxiway surface markings and improved signage would help greatly in reducing the current number of runway incursions.

Low-tech solutions are immediately available without spending millions of dollars on systems development and implementation. They are also available *now*—not many years down the road.

The FAA has even admitted to never fully understanding what causes pilots, controllers, and others to allow runway incursions. Further, the FAA has also admitted that there has been too much reliance on technology to control a human problem. The FAA states that advanced technology will not eliminate runway incursions—education is the key.

Education for Controllers

The FAA has developed several training programs for controllers and pilots designed to make each more aware of the runway incursion problem. These programs are available on a nationwide basis. In some cases, local Flight Safety District Offices (FSDOs) also have designed programs to assist local pilots in becoming more aware of runway incursions and to avoid them.

Monthly Refresher Training

To ensure that air traffic controllers gain and maintain a high level of runway incursion prevention awareness, the FAA has mandated that runway incursion prevention be included in the monthly refresher training at every control tower. Monthly refresher training is provided at the local level and generally includes information from local, regional, and headquarters sources.

Computer Course Instruction

A computer-based instruction (CBI) course entitled, "Preventing Runway Incursions," was mandated for all controllers to be completed by June 2000. It is a four-part course developed to heighten controller runway incursion awareness as well as to enhance overall visual airport scanning and vigilance.

Standardization

To provide standardization at heavily trafficked airports, the FAA, in Standardized Taxi Routes (STR), FAA Order 7110.116, has prescribed standards and procedures for use by airports, air traffic control, and flight standards in the development and utilization of STRs. Simply stated, all the aircraft will be using the same taxi paths under the same circumstances—not scattered at the whims of controllers and pilot requests.

Education for Pilots

The FAA has been promoting runway incursion awareness through the use of pamphlets and posters—and a Web site at *http://www.faa.gov/runwaysafety/* and through many locally developed and presented educational programs at many FSDOs around the country. For current information about FAA-sponsored programs in your area, contact your local FSDO office. An excellent brochure from the FAA about runway incursions is available online at *http://nasdac.faa.gov/safety_products/runwayincursion.htm.*

Enter the AOPA

The Aircraft Pilots and Owners Association (AOPA) has been at the forefront of runway incursion prevention for several years. Unusual for any organization is the public sharing of its information relative to incursions.

The AOPA's Air Safety Foundation operates a very good Web site at *http://www.aopa.org/asf/runway_safety/* that contains an interactive program designed to teach pilots about runway incursion avoidance. The program is divided into three modules.

The Air Safety Foundation went to a lot of effort to produce this interactive educational experience, and it is followed by a short quiz. You can partake of the offering from home—where it is safe to make mistakes.

Enforcement

The FAA says little at this time about enforcement activities, although it is a very good tool to use for runway incursion prevention. When education does not work, a heavy-handed smack in the wallet may. Having your license to fly suspended for a few months will no doubt garner your undivided attention. And even if you do not receive a suspension, you will have spent considerable time and money fighting the FAA.

If you are flying for a living, you get no pay when you do not fly, and you do not fly when your license is suspended. If you wish to fly for a living, a suspension on your record may well preclude you from realizing your dream.

When you get right down to the nitty gritty of runway incursions, they are not legally very different from passing a stop sign or stop light without stopping. And what happens when you get caught doing that? You receive a ticket from some police officer, have to appear in court, and/or have to spend some money for an attorney, a fine, or both.

The point is: If you do not commit a violation, you will not get hurt—financially or physically.

Airport Design

An important factor in reducing runway incursions is the physical design or layout of an airport. Most of today's U.S. airports were designed over 50 years ago, and current improvements consist of only crowding more and more runways and taxiways into the original small area. Neither old airports nor new improvements will culminate in a design that can lower or eliminate runways incursions. In fact, most additions only complicate matters—making them worse for pilots and controllers. Airports with these antiquated runway layouts are rife with runway and taxiway crossings. Washington/Ronald Reagan National Airport (DCA) is just such an airport—an airport where you cannot "get there from here" without crossing something (FIG. 8-1).

It is, however, quite possible to build an airport that is of such design that actual runway incursions are rare.

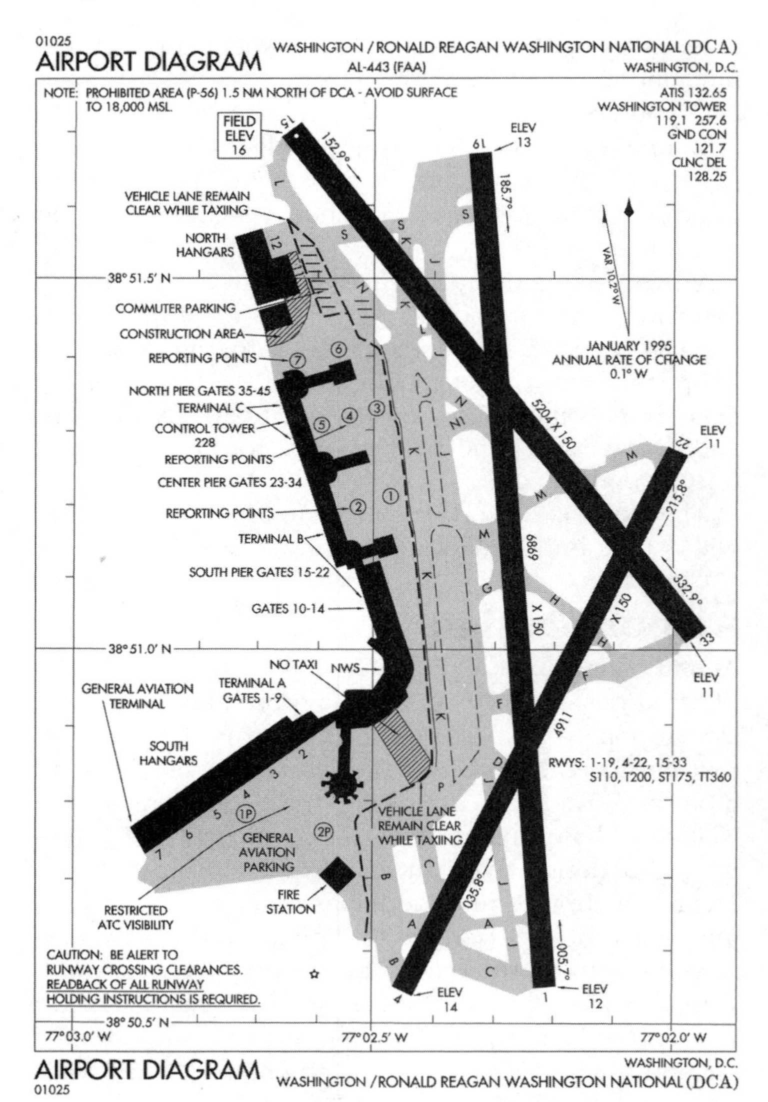

8-1 *Washington/Ronald Reagan Nation Airport (DCA) was designed and constructed prior to World War II. Notice the pattern of runways and taxiways—everything is crossed in every direction (this airport diagram is not suitable for navigational purposes).*

Dulles International Airport (IAD), built in the 1960s, has one of the best runway safety records in the country from 1997 through 2000. Only four minor category D runway incursions were reported by the FAA—technical violations with no chance of a collision. The major reason for Dulles's good safety record is that none of its runways intersect, and aircraft do not have to cross any runways while taxiing. All taxiing is either to or from a runway (FIG. 8-2).

Unfortunately, not all airports with nonintersecting runways and taxiways are free of runway incursion problems. For example, Los Angeles International (LAX), which had the worst 4-year record in the country, has no intersecting runways. The runway and taxiway layout of LAX (FIG. 8-3) is such that many takeoff and landing operations require the crossing of another runway that is also in use.

Where Does All This Leave Us?

All the technology, airport improvements, education, and enforcement still will leave us with some runway incursions. It is doubtful if we will ever truly eliminate all runway incursions—for they are all products of human error.

Remember, "To err is to be human!"

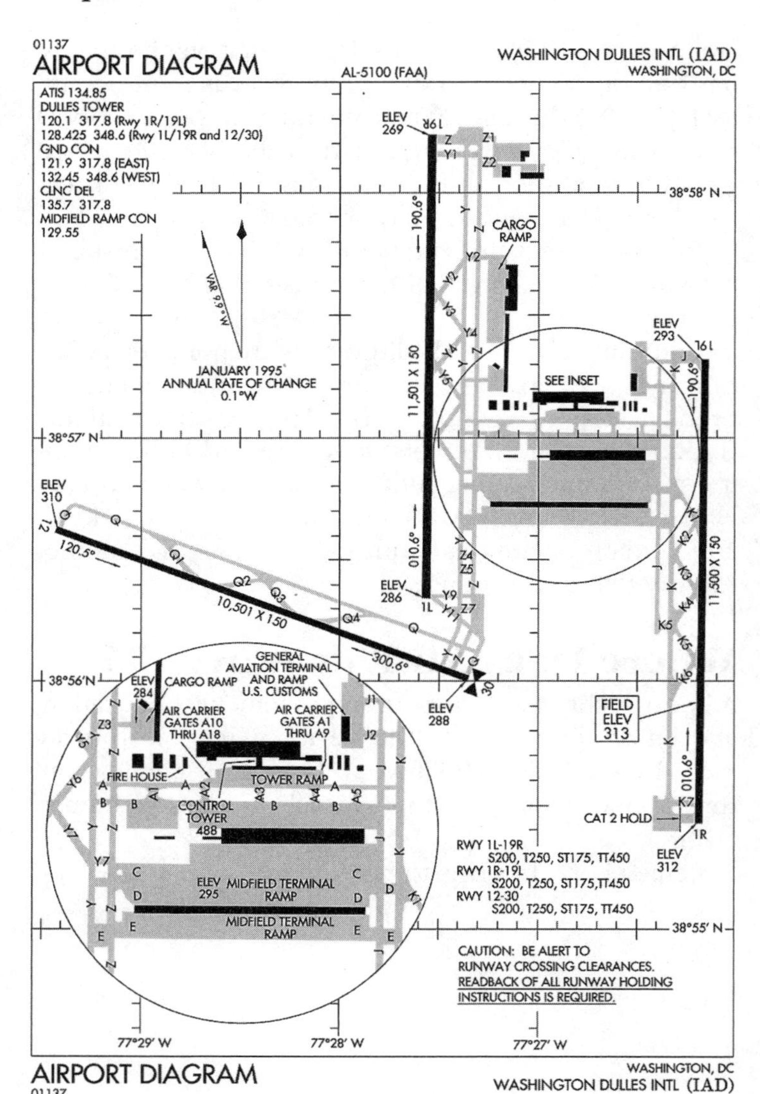

8-2 *Washington Dulles International Airport (IAD) was built with discrete runways. Notice that there are no taxiways crossing any runways (this airport diagram is not suitable for navigational purposes).*

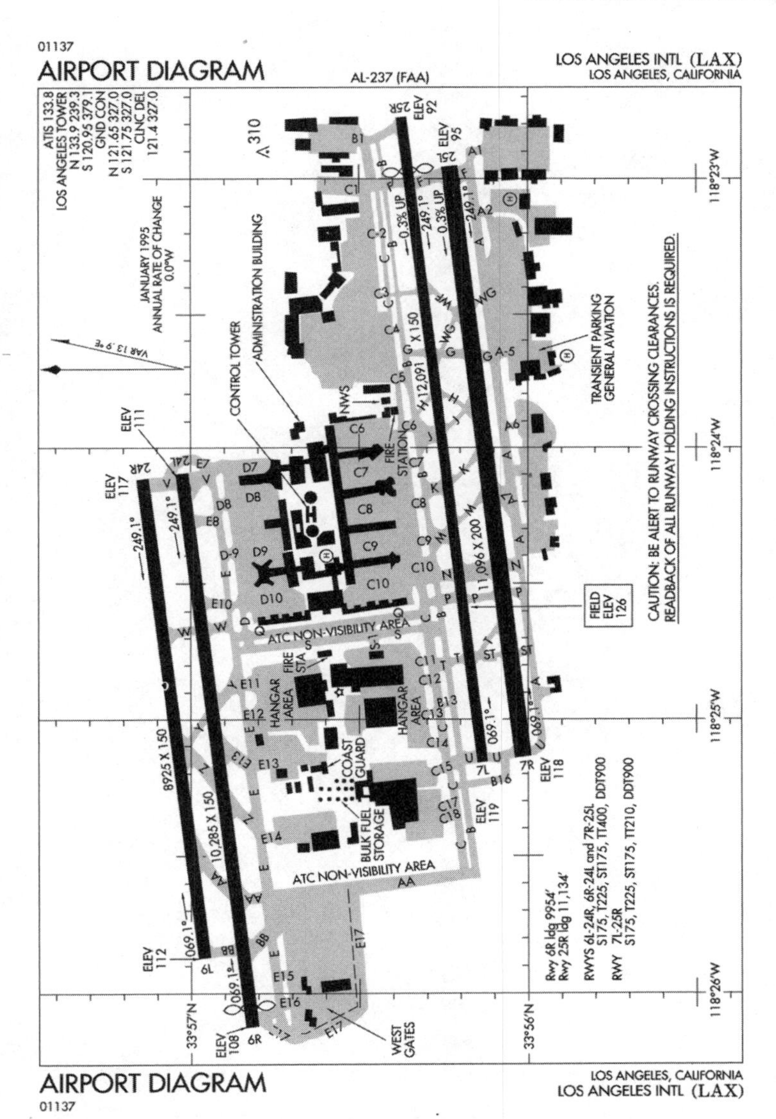

8-3 *Los Angeles International Airport (LAX) has four parallel runways, three of which have taxiways crossing them (this airport diagram is not suitable for navigational purposes).*

Appendix A

Pilot/ATC Glossary

The following is an abridgement of the Pilot/Controller Glossary found in the *Aeronautical Information Manual* (AIM). It will serve to address the most common terms and their usage as are appropriate to avoiding runway incursions:

Abeam An aircraft is *abeam* a fix, point, or object when that fix, point, or object is approximately 90 degrees to the right or left of the aircraft track. Abeam indicates a general position rather than a precise point.

Abort To terminate a preplanned aircraft maneuver; e.g., an aborted take off.

Acknowledge Let me know that you have received my message.

Advise intentions Tell me what you plan to do.

Affirmative Yes.

Apron A defined area on an airport or heliport intended to accommodate aircraft for purposes of loading or unloading passengers or cargo, refueling, parking, or maintenance. With regard to seaplanes, a ramp is used for access to the apron from the water.

Automatic Terminal Information Service (ATIS) The continuous broadcast of recorded non-control information in selected terminal areas. Its purpose is to improve controller effectiveness and to relieve frequency congestion by automating the repetitive transmission of essential but routine information; e.g., "Los Angeles information Alfa. One three zero zero Coordinated Universal Time. Weather, measured ceiling two thousand overcast, visibility three, haze, smoke, temperature seven one, dew point five seven, wind two five zero at five, altimeter two niner niner six. I-L-S Runway Two Five Left approach in use, Runway Two Five Right closed, advise you have Alfa."

Back-taxi A term used by air traffic controllers to taxi an aircraft on the runway opposite to the traffic flow. The aircraft may be instructed to back-taxi to the beginning of the runway or at some point before reaching the runway end for the purpose of departure or to exit the runway.

Blind transmission (See *Transmitting in the blind.*)

Blocked Phraseology used to indicate that a radio transmission has been distorted or interrupted due to multiple simultaneous radio transmissions.

Clear of the runway (a) A taxiing aircraft, which is approaching a runway, is clear of the runway when all parts of the aircraft are held short of the applicable holding position marking. (b) A pilot or controller may consider an aircraft, which is exiting or crossing a runway, to be clear of the runway when all parts of the aircraft are beyond the runway edge and there is no ATC [air traffic control] restriction to its continued movement beyond the applicable holding position marking. (c) Pilots and controllers shall exercise good judgement to ensure that adequate separation exists between all aircraft on runways and taxiways at airports with inadequate runway edge lines or holding position markings.

Cleared for Takeoff ATC authorization for an aircraft to depart. It is predicated on known traffic and known physical airport conditions.

Cleared to land ATC authorization for an aircraft to land. It is predicated on known traffic and known physical airport conditions.

Correction An error has been made in the transmission and the correct version follows.

Emergency A distress or an urgency condition.

Expedite Used by ATC when prompt compliance is required to avoid the development of an imminent situation. Expedite climb/descent normally indicates to a pilot that the approximate best rate of climb/descent should be used without requiring an exceptional change in aircraft handling characteristics.

Final Commonly used to mean that an aircraft is on the final approach course or is aligned with a landing area.

Go ahead Proceed with your message. Not to be used for any other purpose.

Go around Instructions for a pilot to abandon his [or her] approach to landing. Additional instructions may follow. Unless otherwise advised by ATC, a VFR aircraft or an aircraft conducting visual approach should overfly the runway while climbing to traffic pattern altitude and enter the traffic pattern via the crosswind leg. A pilot on an IFR flight plan making an instrument approach should execute the published missed approach procedure or proceed as instructed by ATC; e.g., "Go around" (additional instructions if required).

Hold-short position marking The painted runway marking located at the hold-short point on all LAHSO runways.

Hold-short position lights Flashing in-pavement white lights located at specified hold-short points.
Hold-short position signs Red and white holding position signs located alongside the hold-short point.
How do you hear me? A question relating to the quality of the transmission or to determine how well the transmission is being received.
Immediately Used by ATC or pilots when such action compliance is required to avoid an imminent situation.
I say again The message will be repeated.
LAHSO An acronym for "Land and Hold Short Operation." These operations include landing and holding short of an intersecting runway, a taxiway, a predetermined point, or an approach/departure flightpath.
Land and hold short operations Operations which include simultaneous take offs and landings and/or simultaneous landings when a landing aircraft is able and is instructed by the controller to hold-short of the intersecting runway/taxiway or designated hold-short point. Pilots are expected to promptly inform the controller if the hold short clearance cannot be accepted.
Landing roll The distance from the point of touchdown to the point where the aircraft can be brought to a stop or exit the runway.
Light gun A handheld directional light signaling device which emits a brilliant narrow beam of white, green, or red light as selected by the tower controller. The color and type of light transmitted can be used to approve or disapprove anticipated pilot actions where radio communication is not available. The light gun is used for controlling traffic operating in the vicinity of the airport and on the airport movement area.
Lost communications Loss of the ability to communicate by radio. Aircraft are sometimes referred to as

NORAD (No Radio). Standard pilot procedures are specified in Part 91.

Make short approach Used by ATC to inform a pilot to alter his [or her] traffic pattern so as to make a short final approach.

Monitor (When used with communication transfer) listen on a specific frequency and stand by for instructions. Under normal circumstances do not establish communications.

Negative "No," or "permission not granted," or "that is not correct."

NORDO See *Lost communications.*

Notices to airmen publication A publication issued every 28 days, designed primarily for the pilot, which contains current NOTAM information considered essential to the safety of flight as well as supplemental data to other aeronautical publications. The contraction NTAP is used in NOTAM text.

Out The conversation is ended and no response is expected.

Over My transmission is ended; I expect a response.

Pilot's discretion When used in conjunction with altitude assignments, means that ATC has offered the pilot the option of starting climb or descent whenever he [or she] wishes and conducting the climb or descent at any rate he [or she] wishes. [The pilot] may temporarily level off at any intermediate altitude. However, once [the pilot] has vacated an altitude, he [or she] may not return to that altitude.

Read back Repeat my message back to me.

Report Used to instruct pilots to advise ATC of specified information; e.g., "Report passing Hamilton VOR."

Roger I have received all of your last transmission. It should not be used to answer a question requiring a yes or a no answer.

Say again Used to request a repeat of the last transmission. Usually specifies transmission or portion thereof not understood or received; e.g., "Say again all after ABRAM VOR."

Speak slower Used in verbal communications as a request to reduce speech rate.

Stand by Means the controller or pilot must pause for a few seconds, usually to attend to other duties of a higher priority. Also means to wait as in "stand by for clearance." The caller should reestablish contact if a delay is lengthy. "Stand by" is not an approval or denial.

Taxi into position and hold Used by ATC to inform a pilot to taxi onto the departure runway in take off position and hold. It is not authorization for take off. It is used when take off clearance cannot be issued immediately because of traffic or other reasons.

That is correct The understanding you have is right.

Transmitting in the blind A transmission from one station to other stations in circumstances where two-way communication cannot be established, but where it is believed that the called stations may be able to receive the transmission.

Unable Indicates inability to comply with a specific instruction, request, or clearance.

Verify Request confirmation of information; e.g., "verify assigned altitude."

When able When used in conjunction with ATC instructions, gives the pilot the latitude to delay compliance until a condition or event has been reconciled. Unlike "pilot discretion," when instructions are prefaced "when able," the pilot is expected to seek the first opportunity to comply. Once a maneuver has been initiated, the pilot is expected to continue until the specifications of the instructions have been met.

"When able," should not be used when expeditious compliance is required.

WILCO I have received your message, understand it, and will comply with it.

Words twice (a) As a request: "Communication is difficult. Please say every phrase twice." (b) As information: "Since communications are difficult, every phrase in this message will be spoken twice."

Appendix B

The FAA Interpretive Rule

14 *Code of Federal Regulations* (CFR) Part 91

Pilot Responsibility for Compliance with Air Traffic Control Clearances and Instructions (as published in the *Federal Register*, Vol.64, No. 62, Thursday, April 1, 1999, Rules and Regulations)

Summary

Pilots operating in areas in which air traffic control is exercised are required by regulation to comply with the clearances and instructions of air traffic controllers except in very narrow circumstances. The FAA has consistently construed and enforced this requirement as ascribing to pilots a high level of responsibility to monitor air traffic control communications attentively. Under normal circumstances, the FAA has expected pilots to understand and to comply with clearly transmitted and reasonably phrased clearances and instructions that govern their operations. Nevertheless, a series of recent National Transportation Safety Board (NTSB) enforcement decisions has raised a question regarding the regulatory responsibility of pilots to hear and comply with air traffic control clearances and instructions. This interpretive rule confirms the FAA's historical construction of its regulations that require compliance with air traffic control clearances and instructions.

Effective date: This document is effective March 26, 1999.

For further information contact Eric Harrell, Air Traffic Operations Program, ATO-100, Federal Aviation Administration, 800 Independence Ave, SW, Washington, DC 20591 (202) 267-9155 or James Tegtmeier, Office of the Chief Counsel, AGC-300, Federal Aviation Administration, 800 Independence Ave, SW, Washington, DC 20591 (202) 267-3137.

History

The FAA's general operating and flight rules require pilots to comply with the clearances and instructions of air traffic control, unless they are amended, except in an emergency or in response to a traffic alert and collision avoidance system resolution advisory. Although a number of aviation regulations are based on this requirement, the general responsibility of pilots to comply with air traffic control clearances and instructions is presently located at 14 CFR 91.123(a) and (b). Aviation regulations according the same responsibility as section 91.123 have existed in similar terms for many decades.

As a practical matter, air traffic control communications rely heavily on accurate verbal radio communication. As a result, the FAA has long considered that aviation safety requires air traffic control to function as a cooperative system, in which all participants must share the responsibility for accurate communication. In the FAA's view, the duty of pilots and air traffic controllers alike is adherence to a high standard in communicating clearly, listening attentively, and understanding reasonably.

Bearing in mind these shares responsibilities, when a miscommunication or misunderstanding occurs, the FAA deems responsible the participant who is the initiating or principal cause of the error. For example, the use of unclear terminology, a failure to hear accurately, or a failure to understand a clear transmission can be the initiating or principal cause of a miscommunication. An example in which an air traffic controller's role excuses the pilot might arise from the controller's issuance of an ambiguous clearance or use of misleading terminology that reasonably causes the pilot's misunderstanding. An example in which neither air traffic control not the pilot is to blame for a miscommunication might exist when the aircraft's radio fails.

With respect to the level of attention and comprehension expected of pilots, an interpretation of a regulatory predecessor to 14 CFR 91.123 was published with the regulation from 1955 through 1962. This interpretation reflects an expectation that pilots will pay particular attention to the transmissions of air traffic control, because air traffic controllers frequently must issue clearances that differ from those that pilots anticipate.

It is important that pilots pay particular attention to the air traffic clearance and not assume that the route and altitude are the same as requested in the flight plan. It is suggested that pilots make a written record of clearances at the time they are received and verify the clearance with Air Traffic Control if any doubts exist.

This interpretive language captures the general responsibility of pilots to remain attentive to the content of air traffic control transmissions, as well as the duty of pilots to resolve any confusion they perceive by contacting air traffic control. The FAA's codification of the latter aspect of these responsibilities currently appears in 14 CFR 91.123(a), which requires pilots to request clarification in the event that they are uncertain about an air traffic control clearance or instruction.

With respect to the more general duty of pilots to remain attentive to and to comprehend air traffic control transmissions, the FAA considers responsibility to hinge on the circumstances. It is air traffic control's practice not to presume that a pilot has received a clearance or instruction unless the pilot first acknowledges receipt of the radio transmission. When a clearance of instruction is issued and acknowledged that the pilot nevertheless fails to comply with the transmission, the FAA construes its regulation to indicate pilot responsibility where neither air traffic control involvement nor a mechanical problem causes the pilot's lapse. Thus, when air traffic control transmits a clearance or instruction that is properly acknowledged and there is no evidence of radio malfunction or similar interference with receipt, the FAA presumes that the radio transmission is received in the aircraft cockpit based on the pilot's duty to listen attentively to air traffic control transmissions and to construe them reasonably, if a clearance or instruction is reasonable phrased and received in the cockpit, the pilot's failure to hear or to understand is the result of the pilot's negligence.

In reviewing the FAA's enforcement of FAA regulations, NTSB has historically agreed with the FAA's construction of the air traffic control regulations. In *Administrator v. Wolfenbarger,* for example, an NTSB administrative law judge dismissed the FAA's allegation that a pilot did not comply with an air traffic control instruction to stop his aircraft short of the active runway. Noting that the pilot's radios were working and that air traffic control's radio transmissions were being broadcast, the NTSB granted the FAA's appeal.

Whether radio frequencies are misselected, whether a pilot does not hear because his [or her] attention is elsewhere, or whether he hears a transmission but chooses to ignore it, is irrelevant. As the administrator points out, the law judge's construction (that a pilot might excusably miss an air traffic control transmission without reason would lead to avoidance of all [air traffic control]) instruction violations simply by claiming they were not received. Not only is this a strained reading, but it is inconsistent with our prior interpretation of the rule.

Similarly, *Administrator v. Nelson,* the NTSB agreed that the text of an air traffic control clearance supported the conclusion that the pilot did not exercise the high level of care and attention expected of him when he mistakenly took a clearance, because it was directed to another aircraft. Although a portion of the clearance may have been blocked and therefore not received by the pilot, the NTSB found that the pilot should not have construed the clearance to be directed to his aircraft.

Related to the responsibilities of pilots and air traffic controllers in conducting radio communications, the NTSB has added to a pilot's full and complete read back—or verbal repetition—of an air traffic control clearance or instruction offers a level of redundance that reduces the risk of miscommunication. At the same time, the NTSB acknowledged that FAA regulations do not require pilots to give a full and complete read back. The NTSB observed that there is concern that full read backs can lead to the congestion of radio frequencies and in some instances disserve air safety.

Nevertheless, when pilots incorrectly repeat the air traffic control transmissions, the NTSB's apparent preference for full read backs has led to two inconsistent lines of case law. The first line of NTSB reasoning generally accords with the FFA's interpretation of FAA regulations. In these cases, the NTSB concludes that an air traffic controller's failure to identify and to correct a pilot's erroneous read back contributes to the pilot's error and warrants a mitigation of the sanction for the pilot's regulatory violation.

A second line of NTSB decisions, which diverges from the FAA's long standing construction of FAA regulations, suggests that providing a read back will excuse the pilot even if the pilot is the initiating or principal cause of a miscommunication. In *Administrator v. Frohmuth,* the NTSB appeared to base its decision on a finding the air traffic controller initiated and then supported the two pilots' misunderstanding. In language not directly required for its legal conclusion, the NTSB added that the pilot's full read back placed responsibility to correct the error on air traffic control. Regardless, the NTSB acknowledged the importance of pilots' careful attention to air traffic control transmissions and specified that pilots will, as a general rule, be held responsible for their mistakes.

Despite the limiting language in *Fromuth,* the NTSB recast the decision the following year in *Administrator v. Atkins,* developing a line of reasoning that does not hold pilots responsible for the errors that they initiate.

(In *Fromuth*), "we clarified our precedent by explaining that even if a deviation from a clearance is initiated by an inadvertent mistake on the pilot's part, that mistake will be excused and no violation will be found if, after the mistake, the pilot takes that, but for air traffic control, would have exposed the error and allowed for it to be corrected."

The NTSB expanded this reasoning to excuse pilots based on certain partial read backs in its decision in *Administrator v. Roland.* In *Roland,* the NTSB accepted that a pilot, without explanation, did not hear the altitude portion of his clearance, although he correctly read back another portion of the clearance. The NTSB excused the pilot from responsibility despite his failure to provide a full and complete read back, concluding that the air traffic controller should have questioned the pilot about the part of the clearance that the pilot failed to read back.

More recently, in *Administrator v. Merrell,* the NTSB excused a miscommunication for which the pilot was the initiating or principal cause due to an unexplained "error of perception," resulting in the pilot's acceptance of a clearance for another aircraft and a loss of separation between two commercial flights. The NTSB agreed that the pilot's unexplained error caused the miscommunication and also seemingly agreed that there was no prior or subsequent air traffic control contribution to the pilot's error. The NTSB excused the pilot's error based on his read back, although the pilot's read back was blocked by another radio transmission and could not have been received and corrected by air traffic control.

The NTSB line of reasoning originating in *Fromuth* and presently culminating in *Merrell,* in effect substitutes a duty to provide a full or, in some cases, a partial read back for a pilot's duty to listen carefully to and understand reasonably the air traffic control transmissions received in his or her aircraft. The NTSB interpretation does not correspond to the FAA's construction of FAA regulations and requires correction.

Interpretation

The NTSB's *Fromuth*-based line of decisions deviates from an accurate construction of the FAA's regulations governing air traffic control communications. These FAA regulations require pilots to comply with air traffic control clearances and instructions. Contrary to the NTSB's reasoning, pilots do not meet this regulatory imperative by offering a full and complete read back or by taking other action that would tend to expose their error and allow for it to be corrected. Read backs are a redundancy in that they supply a check on the exchange of information transmitted through the actual clearance or instruction. Full and complete read backs can benefit safety when the overall volume of radio communications is relatively light; however, they can be detrimental during the periods of concentrated communications.

Giving a full read back of an air traffic control transmission could result in the mitigation of sanctions for a regulatory violation when the air traffic controller, under the circumstances, reasonably should correct the pilot's error but fails to do so. Accordingly, the FAA may take this factor into consideration in setting the amount of sanction in FAA enforcement orders. However, the simple act of giving a read back does not shift full responsibility to air traffic control and cannot insulate pilots from their primary responsibility under 14 CFR 91.123 and related regulations to listen attentively, to hear accurately, and to construe reasonably in the first instance.

Economic Considerations

This interpretation is not a change to the subject regulation that must undergo the economic analysis prescribed in Executive Order 12866 or the Regulatory Flexibility Act of 1980. It is not "a significant regulatory action" as defined in the Executive Order or the Department of Transportation Regulatory Policy and Procedures. This interpretive rule will not have a significant impact on a substantial number of small entities and will not constitute a barrier to international trade. Because this interpretive rule merely provides the correct interpretation of a regulation as the FAA has enforced it, it does not impose a separate economic impact, and no further economic evaluation is warranted.

Appendix C

AC 90-42F

Advisory Circular (AC) No: 90-42F

Date 5/21/90

Subject Traffic advisory practices at airports without operating control towers

1. **Purpose.** This advisory circular (AC) contains good operating practices and procedures for use when approaching or departing airports without an operating control tower and airports that have control towers operating part time. This AC has been updated to include changes in radio frequencies and phraseology.
2. **Cancellation.** Advisory Circular 90-42E, dated November 23, 1988, is canceled.
3. **References.** The following AC's also contain information applicable to operations at such uncontrolled airports.
   (a) AC 90-66, Recommended Standard Traffic Patterns for Aircraft Operations at Airports Without Operating Control Towers.

(b) AC 150/5340-27A, Air-to-Ground Radio Control of Airport Lighting Systems.

4. **Definitions.**

(a) **Common Traffic Advisory Frequency (CTAF).** A designated frequency for the purpose of carrying out airport advisory practices while operating to or from an airport that does not have a control tower or an airport where the control tower is not operational. The CTAF is normally a UNICOM, MULTICOM, flight service station (FSS) frequency, or a tower frequency. CTAF will be identified in appropriate aeronautical publications.

(b) **UNICOM.** A nongovernment air/ground radio communication station which may provide airport information at public use airports.

(c) **MULTICOM.** A mobile service, not open to public correspondence use, used for essential communications in the conduct of activities performed by or directed from private aircraft.

(d) **Movement area.** The runways, taxiways, and other areas of an airport/heliport which are utilized for taxiing/hover taxiing, air taxiing, take off and landing of aircraft, exclusive of loading ramps, and parking areas.

5. **Discussion.**

(a) In the interest of promoting safety, the Federal Aviation Administration, through its *Airman's Information Manual, Airport Facility Directory, Advisory Circular,* and other publications provides frequency information, good operating practices, and procedures for pilots to use when operating to and from an airport without an operating control tower.

(b) There is no substitute for awareness while in the vicinity of an airport. It is essential that pilots remain alert and look for other traffic and exchange traffic information when approaching or departing an airport without the services of an operating control tower. This is of particular importance since other aircraft may not have communication capability or, in some cases, pilots may not communicate their presence or intentions when operating into or out of such airports. To achieve the greatest degree of safety, it is essential that all radio equipped aircraft transmit/receive on a common frequency identified for the purpose of airport advisories.

(c) The key to communicating at an airport without an operating control tower is selection of the correct common frequency. The CTAF for each airport without an operating control tower is published in appropriate aeronautical information publications. The CTAF for a particular airport can also be obtained by contacting any FSS. Use of the appropriate CTAF, combined with visual alertness and application of the following recommended good operating practices, will enhance safety of flight into and out of all such airports.

(d) There are two ways for pilots to communicate their intentions and obtain airport/traffic information when operating at an airport that does not have an operating tower: by communicating with an FSS that is providing airport advisories on a CTAF or by making a self-announced broadcast on the CTAF.

6. **Recommended Traffic Advisory Practices.** All inbound traffic should continuously monitor and

communicate, as appropriate, on the designated CTAF from a point 10 miles from the airport until clear of the movement area. Departing aircraft should continuously monitor/communicate on the appropriate frequency from startup, during taxi, and until 10 miles from the airport unless the *Federal Aviation Regulations* or local procedures require otherwise.

7. **Airport Advisory Service (AAS) Provided by an FSS.**

   (a) An FSS physically located on an airport may provide airport advisory service (AAS) at an airport that does not have a control tower or where a tower is operated on a part-time basis and the tower is not in operation. The CTAF's for FSS's which provide this service are published in appropriate aeronautical publications.

   (b) An FSS AAS provides pilots with wind direction and velocity, favored or designated runway, altimeter setting, known traffic, Notices to Airmen, airport taxi routes, airport traffic pattern, and instrument approach procedures information. Pilots may receive some or all of these elements depending on the current traffic situation. Some airport managers have specified that under certain wind or other conditions, designated runways are used. Therefore, pilots should advise the FSS of the runway they intend to use. It is important to note that not all aircraft in the vicinity of an airport may be in communication with the FSS.

   (c) In communicating with an FSS on CTAF, establish two-way communications before transmitting outbound/inbound intentions or information. Inbound aircraft should initiate contact approximately 10 miles from the airport. Inbounds should report altitude, aircraft type, and location relative to the airport; should indicate whether landing or overflight; and should request airport advisory. Departing aircraft should, as soon as practicable after departure, contact the FSS and state the aircraft type, full identification number, type of flight planned; i.e., visual flight rules (VFR) or instrument flight rules (IFR), the planned destination or direction of flight, and the requested services desired. Pilots should report before taxiing, before entering the movement area, and before taxiing onto the runway for departure. If communication with a UNICOM is necessary, pilots should do so before entering the movement area or on a separate transceiver. It is essential that aircraft continuously monitor the CTAF within the specified area.

(d) Examples of AAS phraseology:

(1) Inbound: "VERO BEACH RADIO, CENTURION SIX NINER DELTA DELTA ONE ZERO MILES SOUTH, TWO THOUSAND, LANDING VERO BEACH. REQUEST AIRPORT ADVISORY."

(2) Outbound: "VERO BEACH RADIO, CENTURION SIX NINER DELTA DELTA, READY TO TAXI, VFR, DEPARTING TO THE SOUTHWEST. REQUEST AIRPORT ADVISORY."

8. **Information Provided by Aeronautical Advisory Stations (UNICOM).** UNICOM stations may provide pilots, upon request, with weather information, wind direction, the recommended runway, or other necessary information. If the UNICOM frequency is designated as the CTAF, it will be

identified in appropriate aeronautical publications. If wind and weather information are not available, it may be obtainable from nearby airports via Automatic Terminal Information Service or Automated Weather Observing System frequency.

9. **Self-Announce Position and/or Intentions.**
   - (a) General. "Self-announce" is a procedure whereby pilots broadcast their position, intended flight activity or ground operation on the designated CTAF. This procedure is used primarily at airports which do not have a control tower or an FSS on the airport. The self-announce procedure should also be used when a pilot is unable to communicate with the local FSS on the designated CTAF.
   - (b) If an airport has a control tower which is either temporarily closed or operated on a part-time basis and there is no operating FSS on the airport, pilots should use the published CTAF to self-announce position and/or intentions.
   - (c) Where there is no tower, FSS, or UNICOM station on the airport, use MULTICOM frequency 122.9 for self-announce procedures. Such airports will be identified in appropriate aeronautical information publications.
   - (d) *Practice approaches.* Pilots conducting practice instrument approaches should be particularly alert for other aircraft that may be departing in the opposite direction. When conducting any practice approach, regardless of its direction relative to other airport operations, pilots should make announcements on the CTAF as follows:
     - (1) when departing the final approach fix, inbound;

(2) when established on the final approach segment or immediately upon being released by ATC;

(3) on completion or termination of the approach; and

(4) on executing the missed approach procedure.

NOTE: Departing aircraft should always be alert for arrival aircraft that are opposite direction.

10. **UNICOM Communication Procedures.**

(a) In communicating with a UNICOM station, the following practices will help reduce frequency congestion, facilitate a better understanding of pilot intentions, help identify the location of aircraft in the traffic pattern, and enhance safety of flight:

(1) Select the correct CTAF frequency.

(2) State the identification of the UNICOM station you are calling in each transmission.

(3) Speak slowly and distinctly.

(4) Notify the UNICOM station approximately 10 miles from the airport, reporting altitude, aircraft type, aircraft identification, location relative to the airport, and whether landing or overflight. Request wind information and runway in use.

(5) Report on downwind, base, and final approach.

(6) Report leaving the runway.

(b) Examples of UNICOM phraseologies:

(1) Inbound: "FREDERICK UNICOM CESSNA EIGHT ZERO ONE TANGO FOXTROT 10 MILES

SOUTHEAST DESCENDING THROUGH (ALTITUDE) LANDING FREDERICK, REQUEST WIND AND RUNWAY INFORMATION FREDERICK."

"FREDERICK TRAFFIC CESSNA EIGHT ZERO ONE TANGO FOXTROT ENTERING DOWNWARD/BASE/FINAL (AS APPROPRIATE) FOR RUNWAY ONE NINE (FULL STOP/TOUCH-AND-GO) FREDERICK."

"FREDERICK TRAFFIC CESSNA EIGHT ZERO ONE TANGO FOXTROT CLEAR OF RUNWAY ONE NINE FREDERICK."

(2) Outbound: "FREDERICK UNICOM CESSNA EIGHT ZERO ONE TANGO FOXTROT (LOCATION ON AIRPORT) TAXIING TO RUNWAY ONE NINE, REQUEST WIND AND TRAFFIC INFORMATION FREDERICK."

"FREDERICK TRAFFIC CESSNA EIGHT ZERO ONE TANGO FOXTROT DEPARTING RUNWAY ONE NINE. REMAINING IN THE PATTERN OR DEPARTING THE PATTERN TO THE (DIRECTION) (AS APPROPRIATE) FREDERICK."

11. **Examples of Self-Announce Phraseologies.** It should be noted that aircraft operating to or from another nearby airport may be making self-announce broadcasts on the same UNICOM or MULTICOM frequency. To help identify one airport from another, the airport name should be spoken at the beginning and end of each self-announce transmission.

(a) Inbound: "STRAWN TRAFFIC, APACHE TWO TWO FIVE ZULU, (POSITION), (ALTITUDE), (DESCENDING) OR ENTERING DOWNWIND/BASE/FINAL (AS APPROPRIATE) RUNWAY ONE SEVEN FULL STOP, TOUCH-AND-GO, STRAWN."

"STRAWN TRAFFIC APACHE TWO TWO FIVE ZULU CLEAR OF RUNWAY ONE SEVEN STRAWN."

(b) Outbound: "STRAWN TRAFFIC, QUEENAIRE SEVEN ONE FIVE FIVE BRAVO (LOCATION ON AIRPORT) TAXIING TO RUNWAY TWO SIX STRAWN."

"STRAWN TRAFFIC, QUEENAIRE SEVEN ONE FIVE FIVE BRAVO DEPARTING RUNWAY TWO SIX. DEPARTING THE PATTERN TO THE (DIRECTION), CLIMBING TO (ALTITUDE) STRAWN."

(c) Practice Instrument Approach: "STRAWN TRAFFIC, CESSNA TWO ONE FOUR THREE QUEBEC (NAME—FINAL APPROACH FIX) INBOUND DESCENDING THROUGH (ALTITUDE) PRACTICE (TYPE) APPROACH RUNWAY THREE FIVE STRAWN."

"STRAWN TRAFFIC, CESSNA TWO ONE FOUR THREE QUEBEC PRACTICE (TYPE) APPROACH COMPLETED OR TERMINATED RUNWAY THREE FIVE STRAWN."

12. **Summary of Recommended Communications Procedures.**

(a) UNICOM (no tower of FSS). Communicate with UNICOM station on published CTAF frequency (122.7, 122.8, 122.725, 122.975, or 123.0). If unable to contact UNICOM station, use self-announce procedures on CTAF.

(b) No tower, FSS, or UNICOM. Self-announce on MULTICOM freq. 122.9

(1) Outbound: Before taxiing and before taxiing on the runway for departure.

(2) Inbound: 10 miles out, and entering downwind, and base, and final, and leaving the runway.

(3) Practice instrument approach: Departing final approach fix (name) inbound, and approach completed/terminated.

(c) No tower operation, FSS open

(1) Outbound: Before taxiing and before taxiing on the runway for departure.

(2) Inbound: 10 miles out, and entering downwind, and base, and final, and leaving the runway.

(3) Practice instrument approach: Departing final approach fix (name) inbound, and approach completed/terminated.

(d) FSS closed (no tower)

(1) Outbound: Before taxiing and before taxiing on the runway for departure.

(2) Inbound: 10 miles out, and entering downwind, and base, and final, and leaving the runway.

(3) Practice instrument approach: Departing final approach fix (name) inbound, and approach completed/terminated.

(e) Tower, or FSS not in operation

(1) Outbound: Before taxiing and before taxiing on the runway for departure.

(2) Inbound: 10 miles out, and entering downwind, and base, and final, and leaving the runway.

(3) Practice instrument approach: Departing final approach fix (name) inbound, and approach completed/terminated.

13. **IFR Aircraft.** When operating in accordance with an IFR clearance, if air traffic control (ATC) approves a change to the advisory frequency, change to and monitor the CTAF as soon as possible and follow the recommended traffic advisory procedures.

14. **Ground Vehicle Operation.** Drivers of airport ground vehicles equipped with radios should monitor the CTAF frequency when operating on the airport movement area and remain clear of runways/taxiways being used by aircraft. Radio transmissions from ground vehicles should be confined to safety-related matters.
15. **Radio Control of Airport Lighting Systems.** Whenever possible, the CTAF will be used to control airport lighting systems at airports without operating control towers. This eliminates the need for pilots to change frequencies to turn the lights on and allows a continuous listening watch on a single frequency. The CTAF is published on the instrument approach chart and in other appropriate aeronautical information publications. For further details concerning radio controlled lights, see AC 150/5340-27.
16. **Designated UNICOM/MULTICOM Frequencies.** The following listing depicts appropriate UNICOM and MULTICOM frequency used as designated by the Federal Communications Commission (FCC).

Frequency	Use
122.700	Airports without an operating control tower
122.725	Airports without an operating control tower
122.750*	Air-to-air communications & private airports (not open to the public)
122.800	Airports without an operating control tower
122.900*	(MULTICOM FREQUENCY) Activities of a temporary, seasonal, or emergency nature.
122.925*	(MULTICOM FREQUENCY) Forestry management and fire suppression, fish and game management and protection, and environmental monitoring and protection.
122.950	Airports with control tower or FSS on airport

122.975	Airports without an operating control tower
123.000	Airports without an operating control tower
123.050	Airports without an operating control tower
123.075	Airports without an operating control tower

*UNICOM licensees are encouraged to apply for UNICOM 25 kHz spaced channel frequencies. Due to the extremely limited number of frequencies with 50 kHz channel spacing, 25 kHz channel spacing should be implemented. UNICOM licensees may then request FCC to assign frequencies in accordance with the plan, which FCC will review and consider for approval.

Note 1: In some areas of the country, frequency interference may be encountered from nearby airports using the same UNICOM frequency. Where there is a problem, UNICOM operators are encouraged to develop a "least interference" frequency assignment plan for airports concerned using the frequencies designated for airports without operating control towers.

Note 2: Wind direction and runway information may not be available on UNICOM frequency 122.950.

17. **Use of UNICOM for ATC Purposes.** *UNICOM service may be used for ATC purposes,* only under the following circumstances:
    (a) Revision to proposed departure time.
    (b) Take off, arrival, or flight plan cancellation time.
    (c) ATC clearance, provided arrangements are made between the ATC facility and the UNICOM licensee to handle such messages.
18. **Miscellaneous.** Operations at airports without operating control towers require the highest degree of vigilance on the part of pilots to see and avoid aircraft while operating to or from such airports. Pilots should stay alert at all times, anticipate the unexpected, use the published CTAF frequency, and follow recommended airport advisory practices.

Appendix D

Examples of Airport Signage and Surface Markings

The following examples are designed by the Federal Aviation Administration (FAA) to make pilots aware of the various types of airport surface markings and signage (FIG. D-1).

Mandatory Instruction Signs

Holding Position Signs

Runway

15-33

Beginning of Takeoff Runway

33

ILS Critical Area

Runway Approach Area

Prohibiting Aircraft Entry

Taxiway that Intersects Two Runways

D-1 *Example of signs and runway surface markings (these diagrams are not suitable for navigational purposes).*

Location Signs

Taxiway

Taxiway collocated with a Runway Holding Position Sign

Runway

Runway Boundary Sign

ILS Critical Area Boundary Sign

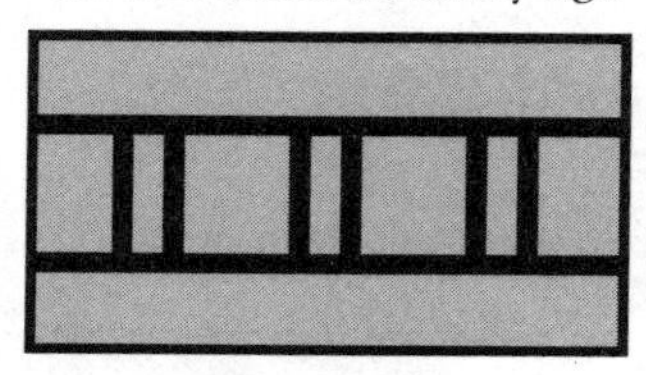

Destination Signs

Military Area

Common Taxi Route to Two Runways

27•33→

Different Taxi Routes to Two Runways

←5 13↑

D-1 (*Continued*).

Direction Signs

Direction Sign Array with Location Sign on the Far Side of the Intersection

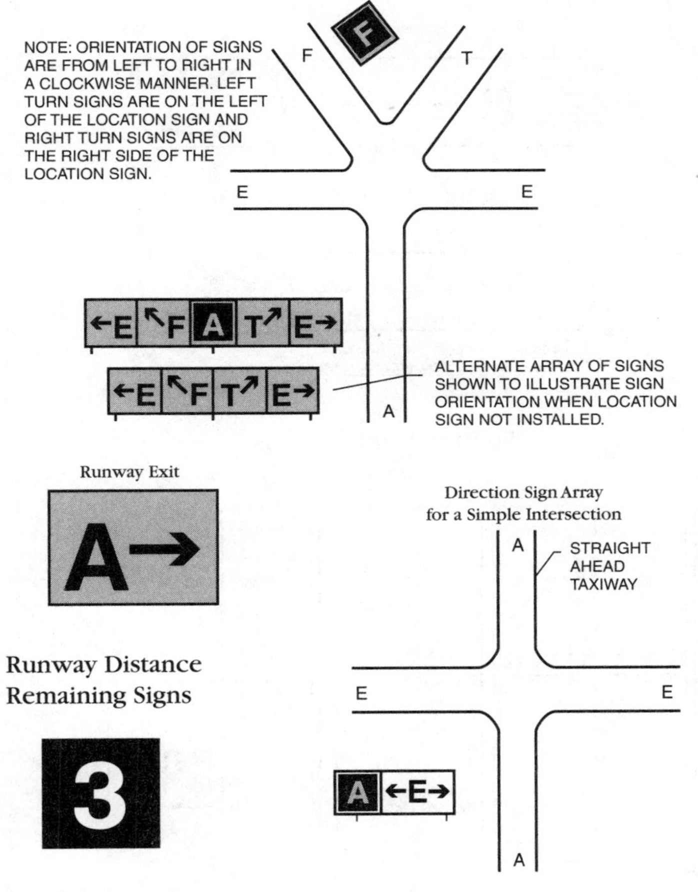

D-1 (*Continued*).

Runway Surface Markings

Nonpercision Instrument Runway and Visual Runway Markings

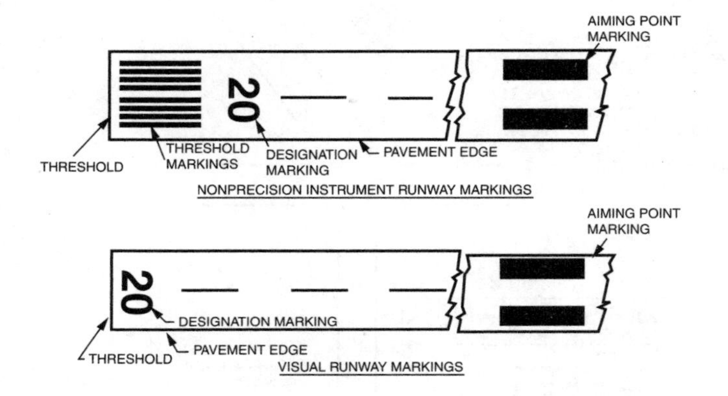

Surface Painted Signs

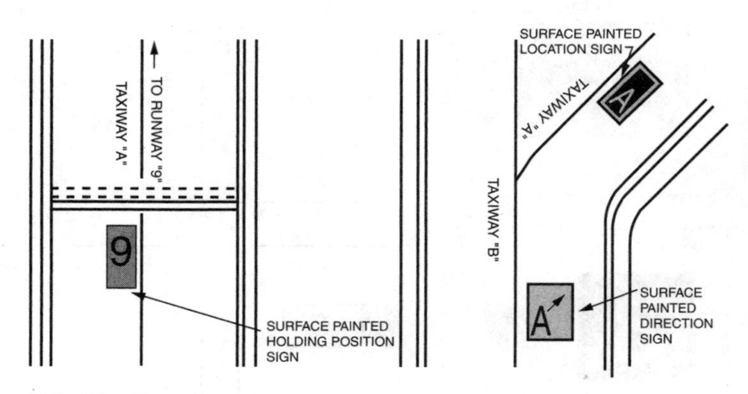

D-1 (*Continued*).

Geographic Position Markings

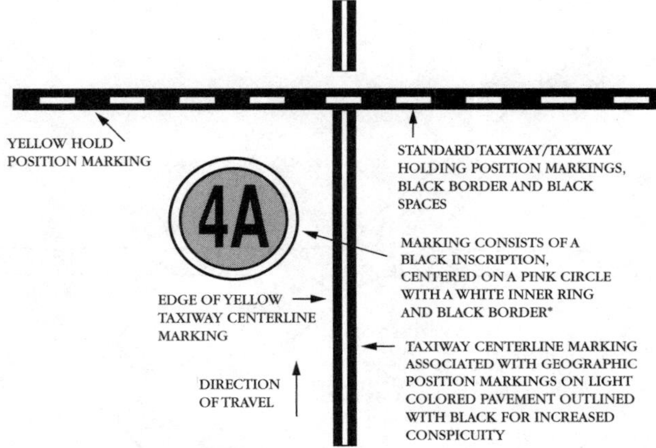

* When installed on asphalt or other dark colored pavements, the white ring and the black ring are reversed, ie., the white ring becomes the outer ring and the black ring becomes the inner ring.

Runway Holding Position Markings

on Taxiways

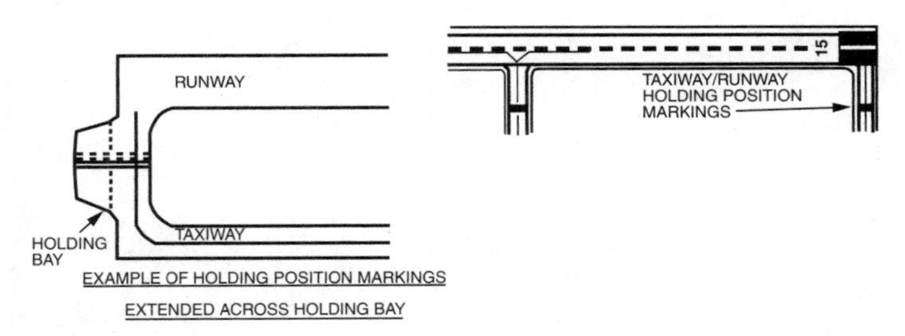

D-1 (*Continued*).

On Runways

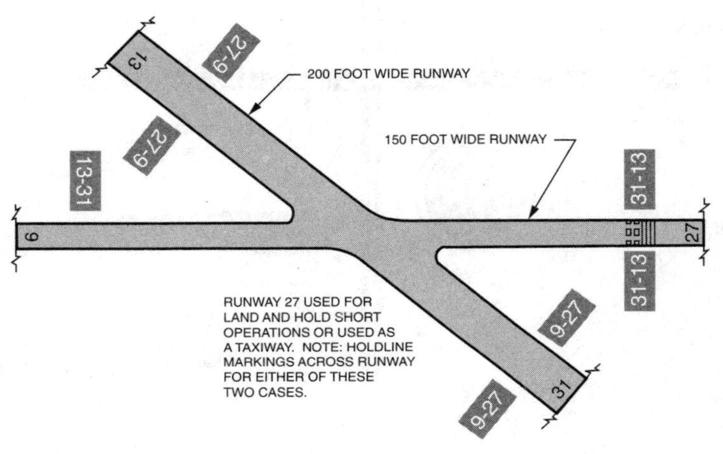

D-1 (*Continued*).

Index

About the Author

Bill Clarke has been a pilot for more than 40 years. Noting the recent increase in the number of runway incursions, he undertook an intense study of the topic and put his findings in this book. He is the author of McGraw-Hill's *A Pilot's Guide to GPS*, 3rd ed. and *The Illustrated Guide to Used Airplanes*, 5th ed.